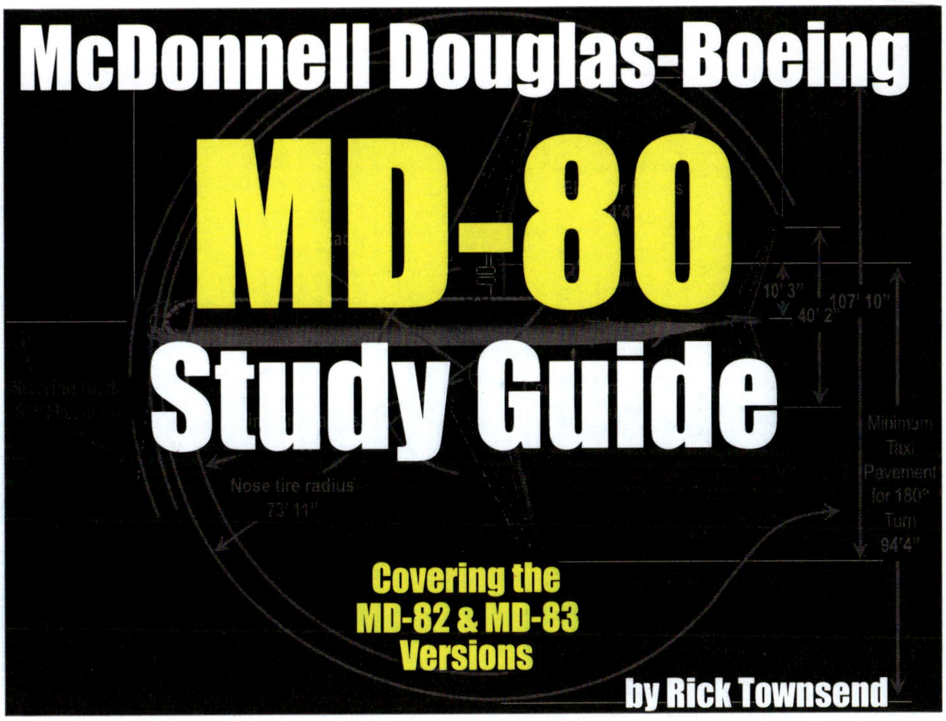

This book was originally developed for pilots at American Airlines. Neither American Airlines nor Boeing Company have endorsed or in any way been responsible for the content of this book.

As a courtesy to the pilots of American Airlines who have supported the development of these books, it is noted that this edition is current to the best ability of the author to make it so, and includes material relevant to:

QRH Revision 17, Temporary Revision — None
Flight Manual Volume I Revision 101, Temporary Revision — None
Flight Manual Volume II Revision 25, Temporary Revision — None
Performance Manual Revision 18, Temporary Revision — None

Note also that as the airplane has now been removed from service, no further updates are anticipated barring customer feedback.

Published by Pilot Study Guides, LLC, Grapevine, Texas.
First Edition and Copyright 1993
Updated continuously (including copyright protection) through the current 2019 edition.

Graphics: With one exception, all graphics were produced by and are the copyright-protected property of the author, except where noted within this edition.

Acknowledgement: Thanks to Norebbo Design Studios, ***www.norebbo.com*** for allowing use of the profile drawing in the ***Airplane General*** section.

Corporation References: American Airlines™ and the Boeing Corporation™, referenced herein, have neither endorsed nor provided any proprietary information for use in these books. Their referencing is for informational awareness only.

ISBN: **978-1-946544-28-5** (Print)
ISBN: **978-1-946544-33-9** (Amazon Kindle)

Table of Contents

Introduction

What This Book IS

This Study Guide is a compilation of notes taken primarily from the flight manual, but also includes elements taken from class notes, computer-based training, and operational experience. It is intended for use by initial qualification crewmembers preparing for orals, and also for systems review prior to recurrent training or check rides. It is assembled in an attempt to organize in one location all the buzz words, acronyms, and numbers the average pilot needs to know in order to get through the events above from an aircraft systems standpoint.

What this Book IS NOT

It is not officially sanctioned by American Airlines, and the author assumes all responsibility for accuracy. (Forward corrections to the address provided, please!) It does not replace study of the operations manual, but instead provides a supplementary source of review material to complement study of official publications. Except for a few specific areas, Except for a few operational issues, it does not include Flight Manual Part 1, or other materials—just airplane systems details, some ideas for organizing the cockpit flow, cold weather operations, etc. The Study Guide is not printed on fancy paper or expensively bound. That makes it easy to tuck into your kit bag or briefcase and study as you travel. Therefore, for these same reasons, it is far less expensive than most commercially produced books of its kind. It is intended to be affordable and usable.

Suggested Uses

A good time to use this book is on layovers. A reasonable plan would be to attempt to review limitations, memorization Red Box items once per sequence, and the rest of the sections once per month. Reviewing sectional study outlines usually leads to trying to remember one of those great plumbing diagrams from the flight manual or synoptics, and so those also get reviewed in the process.

Limitations and Black Border items are laid out so that you may cover the answer with a 3x5 card and quiz yourself. Other sections are in outline form.

A word about acronyms—you will probably see most of the acronyms you have heard of before in this book, as well as a few new ones. Most of us avoid them when possible, using clues on aircraft instrument and systems management panels to jog our memories. But for those you have trouble with, I've included all the ones I could think of. It is not necessary to learn all of them, but only those you need to use to help remember tough sets of items. Thankfully, the fleet management folks believe in keeping these "laundry lists" to a minimum, so you will see fewer of these than you will likely have seen in previous aircraft types.

Unique Formatting Features

Throughout this book, several specific methods are used to highlight information. They include:

Recommended memorization sections

Material highlighted in a light blue box represents topics that are frequently discussed on orals or which simply need to be known without reference to the books. This highlighting is found in systems notes, but not in the limitations section, which all requires a high level of retention. Please let me know of any recommendations you may have for other information that should be included in this category.

Master Caution and **Master Warning** lights are Indicated by highlighted **MC** (Master Caution) and **MW** (Master Warning) highlights within this Guide.

Panel or Warning Light Text—Words or indications found on system panels or switches is indicated by bold, all capital letters such as **LOW PRESSURE**.

Material that applies only to former TWA airplanes is set off in a box like this one, with the TWA code in the lower right corner of the affected box .... **TWA**

Electronic Version

Digital versions have several special hyperlink features. Buttons at the top of each page represent functions available in the iBooks version of this study guide (iPad/smart phone), as well as the Kindle eBook version. To return to the Table of Contents or panel drawings tap the appropriate button at the top of each page. See web site for further details.

Table of Contents	Overhead Panel	Fwd Panel Ctr Console

The corner boxes on each page identify the book section and are not hyperlinked. **TOC**

Panel drawing pages show graphics which are described in this book. Clicking on or touching the colored graphic panels in electronic versions of this book moves to the page with that graphic and its description.

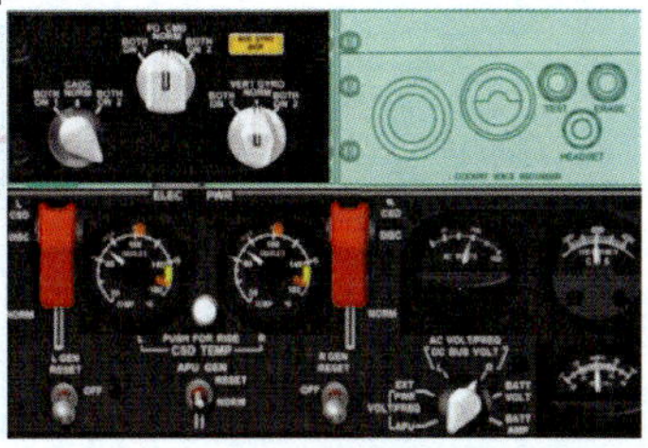

Green shaded areas are not depicted or discussed in this book.

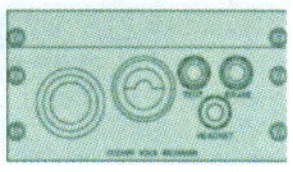

Reproduction

Please don't photocopy this book. I have made a serious effort to keep the costs low for you. In fact, you can buy this book for about half the price of comparable material offered through other commercial publishers. There is a lot of work in this book, as you'll soon be able to see when you study the material. Pages with notes, when provided and annotated as such, may be freely copied by customers. Thanks for your cooperation.

Again, please send me any suggestions or corrections using the contact information below:

Rick Townsend ....Grapevine, Texas
....(817) 416-5533
E-Mail... rick@pilotstudyguide.com
Web Site .... www.pilotstudyguide.com

Revisions

Revisions are made on a regular basis. The downside of revisions is that updating your digital version causes you to lose highlighting and notes you have made. This is because each version includes a change to the printing date, so each page is updated on every new release. The methods for updating digital versions varies by platform. See the Pilot Study Guide web site for details.

To contact the Flight Academy Gift Shop and have a book mailed to your home, you may use the following:

Toll-Free...800-874-1204
DFW Area..817-967-5540
Internet .... www.crewoutfitters.com

Publication Details

Compiled: December 26, 2019

Published by
Pilot Study Guides, LLC, Grapevine, Texas

For ISBN and additional publication details, see title page

Panels

EFIS MAINTENANCE PANEL
AIRCRAFT OPTIONS REGISTER
AIRCRAFT CHECK REGISTER
EMERGENCY FLOOD LIGHT
GROUND PROXIMITY WARNING SPEAKER ASSEMBLY
ENG FIRE DETECT SYS
LOOP A
LOOP B
APU LOOPS
L ENG LOOPS
R ENG LOOPS
BOTH
GROUND SERVICE ELEC PWR
APU PWR
EXT PWR
MAINTENANCE INTERPHONE
OXYGEN REGULATOR
STATUS / TEST
FLIGHT FAULT REVIEW
1-VHF-2-VHF-3
PA
INT
HF-1
HF-2
1-NAV-2
MKR
1-ADF-2
RADIO
VOICE
IDENT
FLT RCDR
NORM
GND TEST
FD CMD
CADC
VERT GYRO
AUX GYRO INOP
COCKPIT VOICE RECORDER
TEST
ERASE
HEADSET
AIR CONDITIONING RECIRCULATION FAN
OFF
ON
AUTO
ENG SYNC
WINDSH
ELEC PWR
L CSD DISC
R CSD DISC
PUSH FOR RISE CSD TEMP
L GEN RESET
R GEN RESET
APU GEN
AC VOLT/FREQ DC BUS VOLT
EXT PWR VOLT/FREQ
APU
BATT VOLT
BATT AMP
CARGO FIRE
CKT BKR LT
STBY COMP LT
OVHD CONSOLE LTS
PANEL
FLOOD
THNDRSTRM LT
CKPT FLOOD
ALT THNDRSTRM
PA VOL
ANTI-SKID
ARM
TEST CKT
STALL TEST
MAX SPD WARN TEST
LOGO LT
YAW DAMP
MACH TRIM COMP
APU L BUS
APU R BUS
L AC BUS
R AC BUS
EXT PWR L BUS
EXT PWR R BUS
GALLEY
AC BUS X TIE
DC BUS X TIE
CLOSE
OPEN
FIRE AGENT
APU
AIR
APU EGT
AIR COND COOLER
MASTER
DOORS AUTO
EMER PWR
FLT DK DOOR
BATT
AIR CONDITIONING
PRESS
L VALVE
R VALVE
TEMP SEL
CABIN SPLY
CABIN
RADIO RACK VENTURI
FAN
CABIN TEMP AUTO
SUPPLY
HP BLD OFF
COLD
HOT
STOP
MANUAL
ENG
START PUMP
CONTIN
START
FUEL HEAT
PNEU PRESS
FUEL TANKS
AFT PUMPS
FWD PUMPS
AFT AUX TRANS
FWD AUX TRANS
STBY
PRIMARY
LDG ALT
LDG BARO
RATE LIMIT
DIFF PRESS
AIR COND SHUTOFF
OVRD
RAM AIR
ICE PROTECT
AIRFOIL
L SYS
R SYS
WINDSHIELD
ANTI-FOG
ANTI-ICE
METER SEL & HEAT
HEATER CUR
AUX
F/O
PITOT
CAPT
RUD TRIM
RAT PROBE
STALL PROBE
STATIC
TAIL
L WINDSHIELD WIPER
R WINDSHIELD WIPER
PARK
SLOW
FAST
ANNUN / DIGITAL LTS TEST
PULL TO DIM

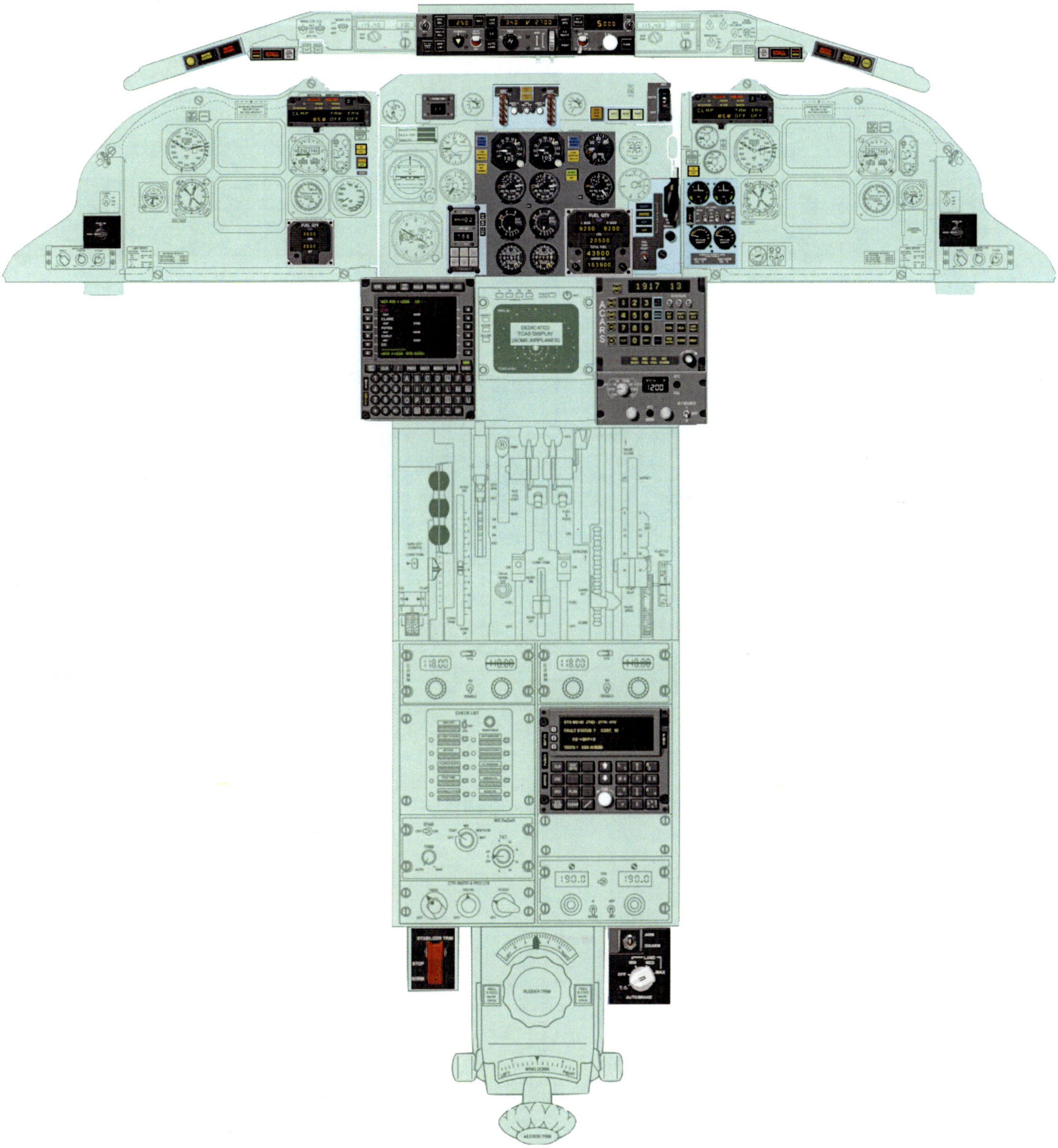

Notes:

- Areas in tinted green are not depicted graphically in this book, but are shown for orientation.
- Electronic version: Touch the colored panel to jump to the page with that system and panel description.

Limitations

This Section is laid out for study with system areas on one side and the appropriate limitation on the other, so you can cover the answer with an index card and quiz yourself. Some of the limitations in Section 2 are not often encountered, such as alternate fuel limitations. For that reason, they have been shortened here for study. These will be listed in the left column, so that as you study, you will be reminded the limitation exists in the book and will (hopefully) think to look there should the situation be encountered. The Limitation column contains only a reminder to check the Ops Manual for the item in question.

Items formatted with green background are from other sections of the operating manual or flight manual Part 1. They are included in this section for ease of study. References for these items are provided.

Items outlined in a dashed box like this are ones which are required to be memorized. The rest need only be learned to a familiarity level.

General Limitations

Instrument Limit Markings (AFM)

Maximum and Minimum Limits:	Red Radial Line
Maximum Limits for Normal Takeoff (Engine N_1, N_2, and EGT):	Orange Radial Line
Precautionary Ranges:	Amber Arc
Normal Operating Ranges:	Green Arc

Engine Display Panel **TWA**

Maximum Limits (N_1, N_2 & EGT Only):	Red Radial / Arc and flashing digits
Maximum Limit Range Normal Takeoff (N1, N2 & EGT Only):	Orange Radial / Arc
Precautionary Range (EGT Only):	Amber Arc
Normal Operating Ranges:	Green Arc

Systems Display Panel **TWA**

Maximum Limits (Oil Temp Only):	Red Annunciator Light
Maximum Limits (Oil Pressure Only):	Flashing digits
Minimum Limits (Oil Pressure Only):	Red Annunciator Light
Precautionary Ranges (Oil Temperature, Pressure, Hydraulic Quantity Only):	Amber Annunciator Light

Wing Landing Light Cooling — After Initial Extension:	1 ½ minutes before next activation (motor cooling)
After Subsequent Extensions:	3 ½ minutes before subsequent activation (motor cooling)
Max Lamp On Time in Ambient, Still Air:	10 Minutes

Operational Limitations

Flight Maneuvering Load Acceleration Limits (AFM)

Flaps Up:	+2.5 to -1.0 G
Flaps Down:	+2.0 to -0.0 G

Operational Limits (AFM)

Takeoff Runway Slope:	+1.7% to -2.0%
Landing Runway Slope:	+2.0% to -2.0%

Maximum Wind Limits—Takeoff and Landing

Tailwind (AFM):	10 Knots (May be further limited by performance requirements)
Crosswind (Max demonstrated):	30 Knots (Maximum demonstrated is not an AFM Limit)

Landing—Crosswind Limits (Including gusts)

Runway (AA Policy)

Dry (Max Demonstrated):	30 Knots
Braking Action Fair:	20 Knots
Braking Action Poor:	10 Knots

Braking action must be determined by ATIS, PIREPs, tower reports or environmental conditions including wet runway, standing water, slush, snow or ice.
Tailwind allowance may be further reduced by performance requirements

Visibility Less than ¾ Mile (RVR 4000' or 1200 Mtrs):	15 Knots — OM Vol. I APP-LDG-G/A, p. 30.5, 35.1, 45.2

Automatic Landings

Headwind:	25 Knots (Reference 1)
Tail Wind:	10 Knots (Reference 1)
Crosswind:	15 Knots (Reference 1)

OM Vol. I APP-LDG-G/A 40.3

Other Wind Limits RUDDER TRAVEL

UNRESTRICTED Light not on during Approach:	12 Knots — QRH, Emergency Sec., FLT-C 16
Maximum Wind Gust:	50 Knots, except in emergency — FM Part 1, p.10-22 & 12-3,& OM Volume I, p. APP-LDG-G/A-45.2
Runway Width < Standard (148'/45 Meters):	20 Knots — FM Part 1, pages 6.1-9
Restricted Captain (FAA Exemption 5549):	10 Knots — FM Part 1, section 10, page 14

Altitudes

Minimum T/O & Landing Altitude (AFM):	-1000 Feet
Maximum T/O & Landing Altitude (AFM):	8500 Feet
T/O & Landing Temperature Limits (AFM):	-54°C to +50°C (-65°F to +122°F)
T/O, Landing, & Enroute Operational Limits:	Environmental Envelope Chart, p. LIM 5 →
Inflight Maximum Altitude:	FL 370

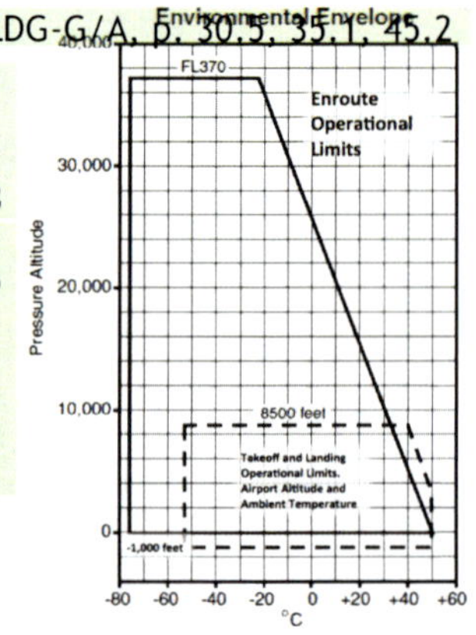

Airspeed

General — Airspeed limitations are given in both airspeed and mach number. *The limiting value is the lower of the two. Altitude determines which value will be lower.*

Maximum Operating Speed V_{MO}: 340 Knots; M_{MO}: .84 M — *Maximum operating limit speed V_{MO} / M_{MO} may not be deliberately exceeded in any regime of flight (climb, cruise, descent).*

Design Maneuvering Speed: Turbulence penetration speed *is a rough approximation. See notes, p. LIM 10.8. No specific numbers identified for memorization*

Landing Gear Operation:

Extension (V_{LO}/M_{LO}):	300 Knots or .70M
Retraction (V_{LO}/M_{LO}):	250 Knots or .70M
Extended (V_{LE}/M_{LE}):	300 Knots or .70M

Note: *The speed limit following use of Alternate Gear Extension is 300 Knots/.70M*

Environmental Envelope

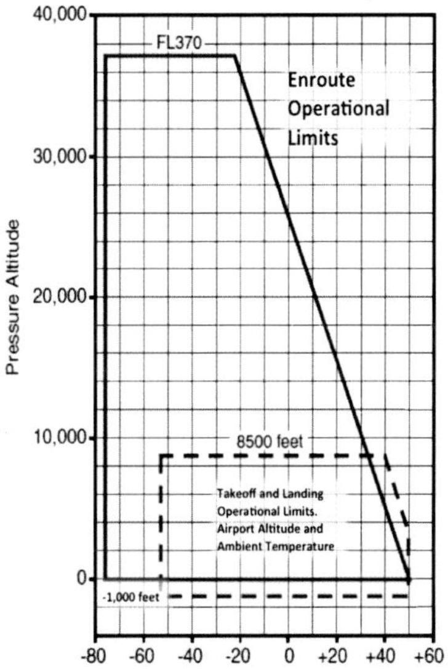

Use of intermediate flap positions, other than 23, (i.e., 18, 24, etc.) is prohibited.

OM Vol. 1, APP-LDG-G/A, p. 10.2

Flap Placard Speeds — (AFM)

V_{FE}:	Max	Min
0-13°:	280 Knots	See Quick-
14-20°:	240 Knots	Reference.
21-25°:	220 Knots	Cards in Cockpit
26-27°:	200 Knots (205, MD-83)	
28-40°:	195 Knots (200, MD-83)	
Limiting Mach Speed M_{FE}:	.57M	

Slats Extended Maximum Speed

Mid Position:	280 Knots/.57M
Full Extension:	240 Knots/.57M

Flap Settings Allowed for Normal Approach & Landing:

Flap Setting	Slats
0	Retracted
0	Extended
11	Extended
15	Extended
23	Extended
28	Extended
40	Extended

Turbulence Penetration Airspeed: The recommended turbulence penetration airspeed is 275 to 285 KIAS or Mach .75 to .79 (whichever is lower). At 10,000 feet and below, minimum recommended speed is 250 KIAS or minimum maneuvering speed (whichever is greater.) Do not fly less than minimum maneuvering speed for existing configuration.

Weight

Maximum Weights	DC-9-82	DC-9-83
Maximum Ramp Weight-(AFM):	150,500 lbs.	161,000 lbs.
Maximum Takeoff Weight-(AFM):	149,500 lbs.	160,000 lbs.
Maximum Landing Weight-(AFM):	130,000 lbs.	130,000 lbs.
Overweight Landings-(AFM):	Landing at any weight that above Maximum Landing Weight is an overweight landing. Refer to FM Part 1, p. 19.1-7 for overweight landing policy, and to QRH - MISC Overweight Landing, p. MISC-0.27	
Maximum Zero Fuel Weight-(AFM):	122,000 lbs.	122,000 lbs.

Air

Air Conditioning Automatic Shutoff Inoperative Dispatch Requirements:	If inoperative, both packs must be off for takeoff.
Cabin Differential Pressure Maximum:	8.07 psi
Maximum, TWA:	7.77 psi
Maximum Emergency Pressure Relief:	8.32 psi
Takeoff & Landing in Manual Control:	Takeoff must be made with Cabin Unpressurized
Unpressurized Flight—Max Altitude following In-Flight Depressurization:	14,000 Feet MSL
Exception—May be exceeded:	When terrain clearance requirements dictate
Maximum Altitude when A/C Dispatched for Unpressurized Flight:	10,000 Feet Pressure Altitude

APU

APU Air Switch:	Off for all in-flight operations.
Engine Starts with APU Supplying Air First Engine:	Air Conditioning Supply Switches OFF
Second Engine:	Air Conditioning supply switch for the operating engine may be on if its pneumatic crossfeed valve is closed.
Operational Altitude Maximum:	Same operational envelope as airplane.
Maximum, TWA:	FL 350

Starter Motor Duty Cycle:	First Attempt:	Must be followed by 5 minutes off	
	Second Attempt:	Must be followed by 5 minutes off	
Third Attempt:		Must be followed by 1 hour (60 minutes) off	
		Note: *Starter duty cycle begins with first indication of RPM.*	

			TWA Airplanes
EGT (AFM)	Starting:	100%	760°C for 30 Seconds
	Max Continuous:	100%	630°C
	Transient:	106%	663°C
Rotor Speed (AFM)	Maximum:	108%	110%

Auto Flight

Auto Throttle Use, with Engine Stall or (AD AD 92-10-163 R1)	Surge on Takeoff:	Must be disconnected
Autopilot Use, General (AFM)	Cockpit Manning Requirements (AFM):	Captain or F/O must be in seat with seat belt on to regain control should autopilot malfunction.
	Coupled Approach with Out-of-Trim Light On (AFM):	No auto coupled approaches if A/P out-of-trim light on more than 3 seconds after the aircraft is stabilized and tracking the glide slope.

Minimum Altitudes for Autopilot Use		Minimum Altitude
Enroute, including Climb & Descent,	Excluding Approaches:	500' AFL
Approaches	Autolandings:	NONE
ILS Coupled Approaches (Other than	Autolandings) IFR:	70 Feet AFL
	VFR:	50 Feet AFL
	NON ILS / RNAV:	50 Feet below MDA

No Automatic Landing Allowed if (AFM) (FLEARRLES):

- ◆**F***light Controls—Any unusual control position or other abnormal conditions exist in the manual flight control system.*
- ◆**E***ither Inertial Reference Unit (IRU) is in Attitude or OFF mode.*
- ◆**A***lign (ALN) Mode is not annunciated on an FMA by 100 feet radio altitude.*
- ◆**R**UDDER CONTROL MANUAL *Light is ON.*
- ◆**R***unway width is less than 145 feet (45 meters).*
- ◆**L***anding overweight*
- ◆**E***ngines—Either Engine becomes inoperative at an altitude greater than 50 feet above the runway.*
- ◆**S***poilers—Automatic Ground Spoilers are not armed and operational.*

Automatic Landings—May be Made under These Conditions

Airplane(LAT):

- ◆**L**MP STATUS *Placard Holder—EMPTY* **OR** *If the MEL 22-90 pocket (right side) is placarded requiring a flight confidence check (FCC)*
- ◆**A***uto-Flight System Operational Requirements for use of AUTOLAND are satisfied. Refer to SYSTEMS, Autoflight, Autopilot Operational Requirements, p. 25.24 - 25.25*
- ◆**T***ested—An Autoflight Pre-Flight Test must be accomplished on any DFGC to be used for an automatic landing on that flight. (AFM; See SYSTEMS, p. 25.1 for test; 25.24-25 for requirements.)*

Runway:

Do not autoland on a runway that the localizer is unusable inside the runway threshold or localizer is unusable for rollout guidance.

Authorization— CAT I, CAT II or CAT III runway (OM Part I p. 10-27 & OM Vol. 1 App-Landing-Go Around 40.2)

- ♦*TCH—Glide slope TCH must be 35' or greater. OM Vol. 1 Approach-Landing-Go Around 40.2)*

Wind(25-15-10-10):

- ♦♦25 *Knots—Maximum Headwind*
- ♦15 *Knots—Maximum Crosswind (may be further reduced by FM Part I, crosswind limitations, and by visibility.)*
- ♦10 *Knots—Maximum Tailwind* *page 6 of this Study Guide*
- ♦10 *Knots—Maximum gust factor, or steady state wind additive over 5 knots. (OM Volume 1, Climb-Cruise-Descent p. 10.3)*

Communications

Boom Microphones:	Required use:	If operative, use is required below 18,000' MSL on aircraft having a placard stating this requirement on the Captain's and F/O's instrument panels.
	MEL Requirements:	MEL 23-12—Boom Microphones; applies if either boom microphone is inoperative.

Electrical

Generator Load		
Engine Generators	Normal:	Less than 1.0
	Normal Maximum:	1.5 for 5 Minutes
	Peak:	Over 1.5 for 5 seconds.
APU Generator	Max on Ground:	1.25
	Maximum below FL 250:	1.0
	Maximum FL 250-300:	0.7
	Maximum above FL 300:	0.625
Generator Limits	Voltage:	115V ± 8V
	Frequency:	400 Hz ± 20 Hz

Battery Limits — Fully Charged, Expected Time:	Fully charged batteries will supply Emergency Power for ≈ 30 minutes
Normal Voltage:	25-33 V
Minimum Voltage EMER PWR Power Selector ON (Battery under load):	25V
Amps Charging:	0-40 Left
Amps—Emergency Power Selector ON:	10-50 Right
DC Bus Limits — Voltage:	22-30V
Load:	1.0
Normal Max Difference Between Indicators:	≤ 0.3
Normal Expected Load Indication:	≥ 0 (Should indicate some load)

Engines, APU

RPM

RPM (AFM)	N_1	N_2
Reserve Thrust:	Red Radial Line (101.6%)	Red Radial Line (102.5%)
Max Thrust:	219—Orange Radial Line (98.8%)	Orange Radial Line (100.9%)

EGT Operating Condition	Temperature	Time
Starting (AFM): Ground:	500°C	Momentary(See Note 1)
Flight:	625°C	Momentary
Takeoff (AFM): Reserve Thrust:	625°C	5 Minutes (See Note 2)
	Red Radial Line (630°C)	2 Minutes
Max Thrust:	590°C	5 Minutes (See Note 2)
	Orange Radial Line (595°C)	2 Minutes
Max Continuous (AFM) / Climb:	Below Amber Arc (580°C)	Continuous
Max Cruise:	540°C	Continuous

Engine EGT Notes:
1. If 500°C EGT limit is exceeded (any duration), immediately shut down engine. Record peak EGT & duration of overtemp in AML. Maintenance action req'd.
2. Use of takeoff thrust (Max and / or Reserve) must not exceed 5 minutes. If any EGT limit (takeoff / in flight) is exceeded, refer to QRH - EMER / ENG / EGT Inoperative or Reads High

Automatic Reserve Thrust (ART) System (AFM): The ART system must be OFF when using the T. O. FLX mode (Standard Thrust) of the thrust rating system.

Oil

Pressure	MD82 / 83	TWA
Normal Pressure:	Green arc	SDP 40-55
Low Pressure:	Amber arc*	SDP 35-40 psi*
Perform Low Pressure Abnormal Procedure:	Below Red Radial Line	SDP < 35 psi**

Tolerable only for remainder of flight, preferably at reduced power setting

***Refer to QRH – Engines*

Note: *Cold weather starts may result in oil pressures outside the green arc(TWA SDP outside green arc—40-55 psi). Engine may not be run above idle until pressure returns to green arc.*

Temperature	MD82 / 83	TWA
Max Continuous:	Less than Amber Arc	SDP 135°C
Max for 15 Minutes:	Amber Arc	SDP 135-165°C

Quantity — Required for Dispatch:	12 Quarts—Flight Duration 4 Hours or Greater 4 Quarts + 2 Qts / Hour per Engine for planned Flight Duration < 4 Hours
Consumption — Allowable Rate:	1 Quart per hour per engine maximum, or E-6 write-up is required. *See LIM 10.17 if exceeded for details.*
Approach Idle — Operations with System Failed to High, One or Both Engines (AFM):	ART must be activated (solenoids actuated) prior to taxi for T/O, and prior to taxi after landing.
Thrust Reversers — Powerback operations:	Not Authorized
In Flight Use:	Do not operate in flight

Power back operational limits remain in the OM. Details have been removed from this Study Guide to eliminate confusion and for space considerations.

Engine Starter Duty Cycle — 1st Start Attempt:	90 Seconds ON, 5 Minutes OFF May Consist of one 90 Second Motoring Cycle or three normal 30-second start cycles in succession. Check that N_2 has decreased to zero between each start attempt.
Subsequent Attempts:	30 Seconds ON, 5 Minutes OFF
Engine Ignition Duty Cycle — CONTIN:	Continuous
OVRD, GND & FLT START:	2 minutes ON, 3 minutes OFF *then* 2 min. ON, 23 min. OFF
OVRD/START Extended Duty Cycle:	20 Minutes ON, 10 Minutes OFF (E-6 Write-up Required)
In Flight CONTIN Inop, OVRD System Used:	OVRD may be operated to extended duty cycle of 20 min. on, 10 min off. AML entry is required when this extended duty cycle is used.
TWA (Most A/C) A/B Ignition Duty Cycle:	Operation of ignition in A, B, BOTH or OVRD is limited to 10 minutes ON, followed by 10 minutes OFF
TWA For Continuous Ignition:	Alternate between A and B positions
Engine Synchronization (AFM):Use Below 1500' AGL:	Must be OFF
Takeoff Thrust ART Inop, Performance Manual Does NOT Authorize STD Thrust:	Use Reserve (RSV) Thrust
ART Inop, Performance Manual Authorizes STD Thrust:	Use Standard (STD) Thrust
ART Operative, Performance Manual Does NOT Authorize STD Thrust:	ART System Armed (ART Switch AUTO, green READY light ON), use MAX Thrust

Flight Controls

Flaps Prohibited Range:	Do not use flap setting ***between*** 13° and 15°.
	13° & 15° ***Are*** authorized settings.
Ground Spoilers Arming-In Flight (AFM):	Do not arm prior to gear extension.
Mach Trim Compensater (AFM)	
System Inop or Malfunctions In Flight:	Switch to OVRD to remove any trim it may be supplying.
Max Speed with Malfunction:	M_{MO}, = .78M
Rudder Power Requirements for Takeoff:	Must be ON for takeoff
RUDDER CONTROL MANUAL Light On In Flight (AFM):	Do not reduce below approach speed OR 135 Knots (whichever is higher) until landing is assured.
RUDDER TRAVEL UNRESTRICTED Light On In Flight (AFM):	If the light stays on >180 KIAS (MD-82) or 200 KIAS (MD-83), then for all operations above these speeds: ◆Rt. Eng Hyd Pump must be **LOW**, and Aux and Power Transfer Pumps **OFF** ***OR*** ◆Rudder Hydraulic Control Lever must be in the **MAN** position.
Slats AUTO SLAT FAIL Light ON (AFM)	
Max Speed with Flap/Slat Handle out of Up/Ret:	240 KIAS
Prior to Slat Extension:	Reduce speed to below 240 KIAS
Slat Position for Takeoff:	Slats must <*verified to*> be at mid position (Flaps 0-13°) or fully extended (Flaps 15-24°) <*prior to*> takeoff. Editor's Note: *<Bracketed wording modified from OM for clarity.>*
Speed Brakes Use in Combination with Flaps & Slats:	Allowed only with Flaps/Slats **UP/RET** ***or*** **0/EXT** configuration
Use in Combination with Landing Gear:	Do not extend or retract gear with speed brakes deployed.
Ground Spoiler Use (Lever beyond EXT):	May not be used in flight (ground use only in this range.)

Flight Instruments

Electronic Flight Instrument System (AFM)	
MAINT CHECK Annunciator Illuminated:	Do not take off with light illuminated.
EFIS Symbol Generator Switch:	Must be in NORM position except for in-flight failures of an EFIS symbol generator.
GFMS Primary Navigation,Approved Display Modes:	Data displayed on EFIS in MAP, or NAV in ROSE, ARC or COMPACT
FD Requirements if MAP Displayed:	For takeoff and landing, FD must be operative on PFD of each pilot who has MAP mode selected on the ND
Use of Compact Mode:	See LIM-10.21

FMS

Conditions of Approval— Program:	Single GFMS system approval for terminal & enroute navigation with program HT9100-005F (or later) installed.
Primary means of Navigation for Routes:	In North America below 70° N Latitude and the Caribbean.
With Single FMS: Approved for Nav:	Approved for terminal and enroute navigation for operations in N. America ***below 70° N*** latitude and the Caribbean.
Limitations When UNABLE RNP is displayed:	♦En Route and terminal IFR operations can be continued providing the system position is ***verified every 15 minutes***, using other approved navigation equipment. ♦Must report degraded navigation performance to ATC.
Use of Dead Reckoning Mode:	♦Navigation cannot be predicated on this mode ***but*** ♦May be used when no other means available
Currency Verification Preflight:	Pilot must verify currency of program before using
Out-of-Date Program Use:	IFR Navigation prohibited unless each selected waypoint and NAVAID is verified for accuracy using current chart data.
Holding— Holding Patterns:	♦GFMS Use authorized
Approaches:	♦GFMS Use on the final approach segment is authorized.
TWA Flight Management System (FMS) (96xx)	
DFGC Version for FMS:	-970 or Subsequent
Use for Range, Fuel Management and Engine Out Terrain Clearance:	Not Authorized. Performance predictions not demonstrated.
Authorized for use with at least two DMEs::	VFR/IFR RNAV operation in accordance with OCEANIC REMOTE, ENROUTE, TERMINAL and NON-PRECISION APPROACH criteria of AC20-130A
Terminal Area and Approach Operation:	Must have one EFIS NAV display in ARC or ROSE mode to crosscheck FMS position with navigation radio data.
Operations requiring RNP Alert Condition:	Must verify position using other available navigation systems
FMS Not Authorized Regions:	N of 80° North, S of 60° South
Raw Data Monitoring Requirements While maneuvering for approach:	One EFIS NAV display must be in the ARC or ROSE mode until established on final approach course to crosscheck FMS position. Note: *When appropriate, compare aircraft position on the map with ILS, VOR, DME, and ADF systems to detect possible map shift errors.*

Vertical and Lateral Navigation	
Minimum Altitude for VNAV Engagement:	1,000' AGL
Minimum Altitude for NAV Engagement:	400' AGL
FMS-VFR Approach:	Authorized in VMC only; may be used as monitor only in IMC
VNAV and NAV use below MDA or DH:	FNAV & NAV must be manually disengaged below MDA or DH
VNAV Use with Autothrottles:	May only be used with both engines operating

Fuel (AFM)

Fuel Management and Loading (AFM)	
Boost Pumps—T/O & Landing:	Two pumps in each tank must be operating. **Exception:** *See MEL*
Aux Tank Pumps—T/O & Landing:	All Aux Tank fuel pumps must be OFF.
Max Fuel Usage, Center Tank, Prior to Aux Tank Fuel Transfer to Center Tank:	No more than 8500 Pounds may be used from center tank prior to transferring all Aux Tank Fuel into the center tank.
Tank Feed Order After Takeoff:	Must use center & aux tanks (if installed) before main tank fuel
Fuel Pump Switch Action when Tank Empties:	Shall be shut off without delay
When Tanks Empty, Related pumps must be:	Shut off without delay
Fuel Density Range:	6.3-7.1 Pounds per Gallon
Max Imbalance, Main Tanks:	1500 lbs.
Aux Tanks:	400 lbs.
Ballast Fuel Limitations:	See Operating Manual, Vol. 1, LIMITATIONS 10.24
Fuel Distribution Exceptions:	For fuel loads where center and aux tank fuel is carried but wing tanks not full, see restrictions: OM Vol. 1, p. LIMITATIONS 10.25
Fuel Specifications Standard Fuels:	Jet A, Jet A-1, JP-5, JP-8
Alternate Fuels:	JP-8 + 100 (Must coordinate with MOC prior to fueling.) Jet A50
Prohibited Fuels:	JP-4, Jet B, Aviation Gasoline (AvGas.) Use of any other fuel except listed above is not authorized unless approved in advance by MOC.
Inadvertent Mixture of Aviation Gasoline (Av Gas)	In the event of inadvertent mixing of aviation gasoline with kerosene during fueling, the associated fuel tank(s) must be defueled.
Center Tank Pump Check (AFM) AD 1989-07-17 — Requirements:	Prior to ***engine start*** on flights where ***center tank fuel is present and will be needed for that route segment***, center tank fuel pumps must be individually checked to verify pump operation. Accomplished by observing both **INLET FUEL LOW PRESSURE** lights out when each individual **CTR** tank pump is activated.
Fuel Pump Circuit Breaker Reset (AFM) AD 2008-11-15 — Inflight:	Prohibited
Ground:	Not permitted until it has been determined by maintenance it is safe to do so

Hydraulics

Hydraulic Pumps Required Operating for All T/Os and Landings: Power Transfer Unit:	ON
Left Engine Hydraulic Pump:	HIGH
Right Engine Hydraulic Pump:	HIGH
Right Auxiliary Hydraulic Pump:	ON
Aux Hydraulic Pump Circuit Breaker Reset (AFM) AD 2008-11-15 — Inflight:	Prohibited
Ground:	Not permitted until it has been determined by maintenance it is safe to do so
Hydraulic Pressure — Engine Pumps High:	2800 to 3100 psi acceptable range
Electric and Engine Pump Maximum:	3200 psi
Engine Pumps Low:	1300 to 1600 psi acceptable range
Engine Pump Low Maximum Pressure:	1700 psi OM Volume 1, SYSTEMS 70.2

Ice & Rain

Ice Requirements: Upper Wing Surface:	***None Allowed*** for takeoff
Lower Wing Surface:	1/8" ***Frost*** max allowed—underwing fuel tank surfaces only.
Upper Wing Inspection Pilot Must verify:	Heater area of the wing is free of ice contamination prior to engine start for takeoff. Verification may be accomplished by assuring over wing heater system **WARM** light is **ON**. **Caution**: *During inspection, include wing surface beyond heater panel, as clear ice may form in these areas.*
Over wing Heater System(AFM) Inspections Requirements — WARM Light ON:	No further action needed
WARM Light Not ON:	Reset by pressing light assembly If **WARM** light comes back on—no upper wing check required If **WARM** light does NOT come back on— Physical (hands on) check of upper wing surface, as described in QRH - ANTI-ICE, RAIN - **Overwing Heater FAIL**, required if either of the following conditions exist: • Ambient temperature is less than 10°C (50°F) and high humidity or visible moisture (rain, drizzle, sleet, snow, fog, etc.) is present. • Frost or ice is present on the lower surface of either main wing fuel tank.

Physical (Hands On)	Note: *See OM Vol. 1 LIM 10.28 and Q*
Upper Wing Surface Check Required:	♦System disarmed ***or*** ♦WARM Light ***not* ON *or*** ♦Either system **FAIL** light **ON** *AND* ♦Temp. less than 10°C(50°F) ***or*** ♦Frost or ice present on either wing main tank lower surface See additional notes, OM Volume 1, LIMITATIONS - 10.28
Icing Conditions Definitions Ground Operations and Takeoff:	RAT 6°C (42°F) or below **AND** Visible moisture present such as clouds, fog with visibility one mile or less, rain, snow, sleet, or ice crystals. *OR* When RAT is 6°C (42°F) or below and the temperature-dew point spread is 3°C (5°F) or less Icing conditions also exist on the ground and for takeoff when the RAT is 6°C (42°F) or below when any form of moisture is present (standing water, snow, slush etc.) which may be ingested into engine inlets, nacelles or sensor probes. Note: *One engine taxi authorized if no significant precipitation is occurring (snow, sleet, freezing rain) that could adhere to or collect in engine inlet.*
In-Flight:	RAT 6°C (42°F) or below **AND** Visible moisture in any form is present (such as clouds , rain, snow, sleet, or ice crystals.)
Engine Anti-Ice Required ON: For Ground, Takeoff and Flight Operations:	When icing conditions exist or are anticipated (as defined above.)
Airfoil Anti-Ice In-flight Use:	When icing conditions exist or are anticipated.
Takeoff Use:	Not used for takeoff or until above 1000' AFL (Airport Analysis based on no use until 1000' AFL.)
Minimum Duct Pressure:	20 psi with airfoil anti-ice on
Minimum Thrust in Icing Conditions:	Sufficient to keep L/R ICE PROTECT TEMP LOW light out
Application of Tail De-Ice is Required:	♦Once every 20 minutes ♦One minute prior to extension of landing flaps ♦After leaving icing conditions, before turning system off
TWA Note, Single Airfoil Anti Ice:	With single Airfoil Anti-Ice Switch, the tail is de-iced automatically for 2½ minutes every 15 minutes and when the system is turned off.
Standing Water and/or Slush (AFM):	♦Engine Ignition must be in CONTIN (TWA A or B) for T/O & landing ♦Static port heaters—ON if temperature ≤ 5°C/40°F
Windshield Heat Requirements (AFM) General:	On and checked for all flight operations
Center, CA, or F/O's Windshield Heat Inop, Max Speed:	315 KIAS under 10,000 feet Pressure Altitude.
Center, CA, or F/O's Outer Glass Ply Cracked:	315 KIAS maximum under 10,000 feet Pressure Altitude. Windshield heat for affected window OFF.
Center, CA, or F/O's Inner Glass Ply Cracked:	235 KIAS maximum under 10,000 feet Pressure Altitude. Windshield heat for affected window OFF. Note: *No speed restriction above 10,000 feet and no speed restrictions associated with clear view or eyebrow windows.*

Landing Gear

Anti Skid Requirement for ALL Ops (AFM):	Must be Operative
Auto Brakes With Hydraulic System Failures:	Do not use Auto-brakes if either hydraulic system fails
Brake Temperature Maximum for T/O:	205°C – Do not take off if any brake temperature exceeds.
Maximum Temperature to set Parking Brake:	300°C
Brake Wear Limits:	With Parking Brakes parked, indicators must extend beyond the brake housing as follows:
MD-82:	Flush or greater
MD-83:	½ Inch above flush or greater
Maximum Tire Speed:	195 Knots
Brake Pressure Bleed Down:	See Flight Manual, LIMITATIONS, p. 10.31

Miscellaneous

Evacuation Systems Slide Arming Required:	Any time airplane is in motion
With Passengers on Board:	At least one slide must be armed
Aft Bulkhead Door Emergency Operating Handle:	Must be exposed with normal handle covered from time airplane moves from gate until it stops moving on arrival
Emergency Lights Must be armed:	For all flight operations
Oxygen Pressure Minimum at 70°F Bottle Temp.:	1100 psi. See OM Vol. 1, LIM 10.32 for other temperatures
Flight Deck Door and Flight Access System Preflight:	Required once each flight day
Access Code Procedures:	See OM Vol. 1, LIM 10.32 for other temperatures
Locking with L DC Bus Unpowered:	Use Deadbolt
Use of Deadbolt Position with Key:	Ground Only

Navigation

Navigation Instrument Tolerances

CA's ND Course Pointer vs. FO's #1 VOR RMI Pointer:	5°
FO's ND Course Pointer vs. CA's #2 VOR RMI Pointer:	5°
#1 ADF/VOR Pointer vs. #2 ADF/VOR Pointer tuned to same station:	8°
CA's ND Compass Card vs. FO's Compass Card:	5°
TWA Slewing Compass After Inflight Alignment:	Controllers should not be slewed (may result in attitude errors)
96xx Airplanes, Alignment Movement:	Aircraft must be stationary (Normal motion due to passenger and baggage loading is within tolerances
Temperature Difference Between Units:	11° (if more than this, **IRS ALIGN HOLD** light
Temperature Minimum:	-20° (if below this, **TEMP LOW** light)
Latitudes for for 2.5-10 minute alignment:	Between 70° N and 60° S
Maximum northern latitude for 15 minute alignment:	78.25° N
Minimum Equipment Required:	One Multifunction Control and display unit (MCDU)

Warning & Alert

Traffic Alert and Collision Avoidance System (TCAS) Authorized Deviation:	To the extent necessary to comply with a TCAS II RA
Initiating Evasive Maneuvers for TA's Based Only on Information Shown on the TCAS Traffic Display:	Prohibited. These displays and advisories are intended for assistance in visually locating the traffic and lack the resolution necessary for evasive maneuvers.
Compliance with an RA:	Required, unless in the opinion of the Captain, doing so would compromise the safe operation of the flight. (Applies even if visual)
Maneuvers in Response to an RA which are in the Opposite Direction of that Advisory:	Prohibited, unless they are the only means to assure safe separation.
Other critical Warnings such as Stall, Windshear or GPWS:	Take precedence over an RA.
Manner of Response:	Manually and smoothly (A/P and A/T off)
Impact of Transponder altitude reporting OFF:	TCAS is disabled
Other Critical Warnings' Precedence:	Warnings such as stall & GPWS take precedence over RA response
Weather Radar—Do Not Operate Hangar:	Within a hangar
Fueling:	Within 50' of fueling operations or fuel spills
Personnel:	Within 160'
Warm Up:	In Standby Position Only (If applicable)
Enhanced GPWS (AFM) Standby Altimeter:	Must be set to QNH
Authorized Deviation:	To the extent necessary to comply with an EGPWS warning
Navigation Based on use of Terrain Display:	Prohibited
Lighting Requirements for use at Night:	Instrument panel flood lights must be operative
Use at airport not contained in EGPWS Database:	Must be inhibited within 15 minutes of takeoff, approach or landing Note: *All authorized AA airports (regular, provisional, refueling, alternate and designated emergency airports) are contained in the database.*
TWA Use with FMS White or Amber **RNP** message in upper left corner of EFIS PFD:	Must be inhibited
Position Updating Requirement:	EGPWS must be inhibited (Terrain **OVRD** Switch to **OVRD**), unless ◆FMS position is updated at the end of the runway, ***or*** ◆Verified with actual runway position, ***by*** ◆◆Ensuring that with the Mode Select Panel Range selected to 10NM, A/C symbol is on the runway symbol at the appropriate end of the runway, ***or*** ◆◆The "Internal GPS card not navigating" message is present.
Overspeed Warning Inoperative, Maximum Speeds— V_{MO}:	325 Knots below 25,300 Feet
M_{MO}:	.79 Mach above 25,300 Feet See additional notes Limitations 10.36 Caution: *If overspeed warning system is deactivated, carefully monitor Mach/Airspeed Indicator(s). When the black* **CAWS FAIL ANN** *circuit breaker is pulled, aural warnings are also inoperative for* ***engine fire*** *and* ***horizontal stabilizer position.***
Windshear Alerting & Guidance System (WAGS) (AFM) AD 1992-03-06 Bank-Related Desensitization:	During sustained banks of greater than 15°, WAGS is desensitized and alerts resulting from encountering windshear conditions will be delayed.
During Single Engine Operation:	WAGS Capability has not been demonstrated for SE operations.

Emergencies

Memory Items

Procedural changes have been made such that there are now only three memory items:

Complete Loss of AC Power (QRH Electrical 6.1)

Condition: Loss of normal AC and DC power has occurred

1 EMER PWR SWITCH....................ON

Reverser Deployed or ENG REVERSE THRUST and/or UNLOCK Illuminated Inflight (QRH Engines 7.3)

Condition: Reverser is deployed or unlatched inflight.

1 Autopilot/Autothrottle....................OFF
2 Throttle (affected engine)....................(Confirm) IDLE

Runaway Stabilizer (QRH Flight Controls 9.1)

Condition: Uncommanded stabilizer trim movement occurs continuously.

Note: Extended trim operation may result in trim motor thermal shutdown. Trim motor operation may return after sufficient cooling period.

1. Autopilot (If engaged)....................Disconnect
2. **If** runaway movement continues:
   Control wheel trim switches....................Trim opposite direction of runaway as necessary

Former Memory Items

Note: *Procedures below this point on this page were formerly memorized due to the time-critical nature of the tasks involved. A policy change now involves reading and doing these items from a Quick Reference Card (QRC) stored in the cockpit. The procedures remain here for reference, but will be removed in a future revision if no further policy changes occur.*

Airspeed Unreliable (QRH Flight Instruments, Displays 10.1)

Condition: Possible malfunction of pitot static system, air data computer, or instruments.

1 Autopilot/Autothrottle/Flight Directors....................OFF
2 Airplane pitch/thrust:
   If flaps &/or slats EXT....................10° & 80% N1
   If flaps and slats UP....................4° & 80% N1
3 METER SEL & HEAT switch....................Verify not OFF

CABIN ALT / Rapid Depressurization (QRH Air Systems 2.1)

Condition: One or more of these occur:
- Cabin altitude exceedance.
- In flight, a modulating cabin altitude warning horn sounds, followed by the words, "*CABIN ALITITUDE*" and **CABIN ALT** light illuminates.

Oxygen masks....................ON / 100%
2 Communications....................Establish
3 CABIN ALT control lever/wheel....................MANUAL/FULL FORWARD
Note: Manual control wheel forces may be high. Apply force as required.
4 PNEU X-FEED VALVE levers....................Close
5 AIR CONDITIONING SUPPLY switches....................AUTO
6 Passenger Oxygen Masks (if required)....................Deploy
If cabin altitude has exceeded 14,000 feet and passenger oxygen masks have not deployed, move PAX OXY MASK switch to EJECT.

Cockpit Smoke Removal – Unpressurized (QRH Fire Protection 8.17)

Condition: Smoke is present in the cockpit.

1. Oxygen masks and goggles....................ON / 100% / EMERGENCY
2. Communications....................Establish

Engine Fire / Damage / Separation (QRH Fire Protection 8.18)

Condition: One or more of these occur:
- Engine fire warning
- Airframe vibrations with abnormal engine indications

At or Above Engine Out Acceleration Altitude:

1. Autothrottle (if engaged)....................OFF
2. Throttle (affected engine)....................(Confirm) IDLE

TAIL COMPT TEMP HIGH (QRH Air Systems 2.3)

Condition: Tail compartment temperature is high.

1. PNEU X-FEED VALVE Levers....................CLOSED
2. AIR FOIL anti-ice switches....................OFF
3. AIR CONDITIONING SUPPLY switches....................HP BLD OFF

Two Engine Flameout (QRH Engines 7.5)

Condition: Both engines have loss of thrust.

1. EMER PWR switch....................ON
2. ENG IGN selector....................OVRD

Maneuvers

Note: Only selected maneuvers are reproduced here for study. Notes are abbreviated for brevity and ease of study. Refer to QRH, MANEUVERS section for other procedures and full text of notes. Procedures are taken from the QRH, which states, **"This section is a consolidation of emergency maneuvers. Pilots are expected to be proficient in the performance of these maneuvers."**

Approach to Stall or Stall Recovery

WARNING

During takeoff, a stick shaker, STALL warning light, horn, or "STALL" aural warning at rotation may indicate an improper flap/slat configuration. PF immediately calls out, "SLATS EXTEND" and PM confirms, "SLATS EXTENDED."

If control column does not immediately provide the needed pitch response, full forward control input may be necessary. Excessive use of pitch trim may aggravate the condition, or may result in loss of control, or high structural loads.

CAUTION

Do not ignore short duration warnings. Take immediate action.

Note

Do not use flight director commands during the recovery.

Premature recovery may result in a secondary stall or inability to accelerate with thrust available.

All recoveries from approach to stall should be done as if an actual stall has occurred. If conditions permit, accept an altitude loss while accelerating to minimum maneuvering speed for existing configuration.

(OM Vol. 1, MANEUVERS 10.1.1, 10.1.2):

Pilot Flying	Pilot Monitoring
First indication of stall (buffet or stick shaker)	
Initiate the recovery: – Hold control column firmly – Disconnect autopilot and autothrottle – Apply takeoff/go-around or MCT thrust as applicable. If ground contact is imminent, apply thrust to mechanical stops. – Smoothly apply nose down elevator to reduce the angle of attack until buffet or stick shaker stops	– Monitor altitude and airspeed – Verify all required actions have been accomplished Call out – Any omissions – Trend toward terrain contact
Continue the recovery: – Roll in the shortest direction to wings level, if needed – Confirm speed brakes stowed – Advance thrust levers as needed – If flaps/slats or landing gear are extended, do not change configuration during stall recovery	– Monitor altitude and airspeed – Verify all required actions have been accomplished Call out – Any omissions – Trend toward terrain contact
Complete the recovery: – Check airspeed and adjust thrust as needed – Establish pitch attitude – Return to the desired flight path – Re-engage the autopilot and autothrottle if desired	– Monitor altitude and airspeed – Verify all required actions have been accomplished Call out – Any omissions – Trend toward terrain contact

Ground Proximity Warning Actions & Callouts

Correct the flight path or aircraft configuration for the following:
- CAUTION OBSTACLE
- CAUTION TERRAIN
- DON'T SINK
- GLIDESLOPE
- TOO LOW FLAPS
- TOO LOW GEAR

"GLIDESLOPE" may be canceled or inhibited if:
- Conducting a localizer or back-course approach
- Circling approach for an ILS
- Conditions require a deliberate approach below glide slope
- Glide slope signal is unreliable

[1] If go-around thrust is exceeded at any point during the escape maneuver, make an AML entry.

Note

If a terrain caution occurs and positive visual verification is made that no obstacle or terrain hazard exists when flying under daylight VMC conditions, the alert may be regarded as cautionary and the approach may be continued.

Some aural warnings repeat.

(OM Vol. 1, MANEUVERS 10.2.6)

Pilot Flying	Pilot Monitoring
Ground proximity warning aural alert activated	
– Disconnect autopilot – Throttles - TOGA or full forward (if ground/obstacle contact is imminent)[1] – Simultaneously roll wings level and rotate to an initial pitch attitude of 20° – Retract speedbrakes – Trade airspeed for climb performance. If necessary (to prevent ground contact), continue to increase pitch attitude until stick shaker actuates.	– Verify power settings – Verify all required actions have been completed
"Set go-around thrust."	
– Do not change gear or flap configuration until terrain separation is assured – Monitor radio altimeter for sustained or increasing terrain separation – When clear of terrain, slowly decrease pitch attitude and accelerate	Call out: – Any omissions – Any trend toward terrain contact – Monitor vertical speed and altitude (radio altitude for terrain clearance and barometric altitude for a minimum safe altitude)

Moderate to Heavy Rain, Hail or Sleet

Flights should be conducted to avoid thunderstorm or hail activity. If visible moisture is present at high altitude, avoid flight over the storm cell. (Storm cells that do not produce visible moisture at high altitude may be overflown safely.) To the maximum extent possible, moderate to heavy rain, hail or sleet should also be avoided.

(OM Vol. 1, MANEUVERS 10.3)

If moderate to heavy rain, hail or sleet is encountered:

ENG IGN switch .... **CONTIN or A or B**
PNEU X-FEED VALVE levers .... **Open**
ENG / AIRFOIL ANTI-ICE switches .... **As required**

Note

Engine and airfoil anti-ice systems should be off if RAT is above 6°C and no icing is encountered or anticipated. Reduced engine bleeds will increase engine flameout margin during periods of heavy water ingestion.

Autothrottle .... **OFF**
Throttle... Adjust slowly

If thrust changes are necessary, move throttles slowly. Avoid changing throttle direction until engines have stabilized at a selected setting. Maintain an increased minimum thrust setting.

Airspeed .... **Use a slower speed**

Using a slower speed improves engine tolerance of heavy precipitation intake.

Consider starting APU.

Rejected Takeoff

CAUTION

Should directional control become a problem while in reverse thrust, reduce reverse thrust to reverse idle (or forward idle, if required), regain directional control and re-apply reverse thrust as necessary.

Note

If an engine remains at high forward thrust, shut it down using fuel lever.

Prior to 80 knots, the takeoff should be rejected for any of the following:
- Activation of the MASTER WARNING/MASTER CAUTION
- System failure(s)
- Unusual noise or vibration
- Tire failure
- Abnormally slow acceleration
- Unsafe takeoff configuration warning
- Fire or fire warning
- Engine fire/engine failure/compressor stall
- Windshear warning
- Aircraft is unsafe or unable to fly

Above 80 knots and prior to V1, the takeoff should be rejected for any of the following:
- Fire or fire warning
- Compressor stall
- Engine failure
- Windshear warning
- Engine fire
- Aircraft is unsafe or unable to fly

See additional notes in Maneuvers section.

(OM Vol. 1, MANEUVERS 10.4.1)

Captain	First Officer
The captain decides to reject the takeoff	
"Reject, my aircraft."	
	If aircraft control is transferred
	"Your aircraft."
Auto Spoilers - ARMED Without delay, rapidly and simultaneously: – Retard throttles to idle – Apply maximum manual braking – Initiate reverse thrust consistent with conditions – Verify spoilers extended. **If** spoilers fail to extend, extend manually Be alert for directional control problems from asymmetrical reverse thrust Auto Spoilers - Not ARMED, or TWA aircraft Without delay, rapidly and simultaneously: – Retard throttles to idle – Apply maximum manual braking – Extend spoilers – Initiate reverse thrust consistent with conditions Be alert for directional control problems from asymmetrical reverse thrust	Apply slight forward pressure on control column. Verify: – Throttles idle – If an engine remains at high forward thrust, immediately advise captain – Spoiler Lever full aft
	"Deployed."
	If spoilers do not deploy (or fail to remain deployed):
	"No spoilers."
	Captain will manually deploy spoilers.
	Reverse thrust applied. If an engine fails to reverse:
	"No reverser left engine." or "No reverser right engine." or "No reversers."
	Advise control tower as soon as practical, especially during low visibility conditions
At 100 knots	
	"100."
At 80 knots	
	"80."
At 60 knots	
	"60."
When the aircraft is stopped	
Consider accomplishing the following: – Advise F/As and passengers to remain seated or to evacuate – Request ARFF – Complete checklist (if appropriate) for conditions which caused the RTO – Clear the runway, if feasible – Call for Maintenance inspection	

Tailstrike

CAUTION

Do not pressurize the aircraft. Pressurizing the aircraft may cause further structural damage.

(OM Vol. 1, MANEUVERS 10.5)

Cabin Altitude Control Lever .... **MANUAL (Down)**
Outflow VALVE .... **Full aft**

Plan to land at the nearest suitable airport. Conditions permitting, land at an company station where inspection / repair can be accomplished.

Engine Failure - Takeoff Actions & Callouts

[1] The FD TAK OFF mode will command the above schedule. If bank angles of more than 15° are used, accelerate to 0/RET MIN MAN speed.

[2] Recommended one engine climb speed is Slat Retract Speed +20 knots.

(OM Vol. 1, MANEUVERS 10.6.2)

1 ENGINE

Pilot Flying	Pilot Monitoring
Engine fails on takeoff	
"My aircraft."	
Maintain directional control	
At V_R	
	"Rotate."
Rotate to takeoff attitude	
After verifying positive rate of clime on VSI	
Verify positive rate of climb on the altimeter	Verify positive rate of climb on the altimeter
	"Positive Rate."
"Gear up." "Runway heading, heading select"	On command: – Position landing gear lever UP – Disarm spoilers
Use the following climb speeds to the TPS engine-out acceleration altitude[1] : – If an engine failure occurs after V1, but not above V2: • Maintain V2 – If an engine failure occurs after V2: • Maintain speed attained at time of failure not to exceed V2 + 10 – If an engine failure occurs at a speed greater than V2 + 10: • Reduce to and maintain V2 + 10	– Monitor engine and flight instruments – Monitor climb speeds
At engine out acceleration altitude or higher (if required)	
"Altitude Hold" "Flaps Up." and "Slats Retract" on schedule	– Monitor engine and flight instruments – Monitor climb speeds On command: – Select ALT HLD – Retract flaps / slats
At slats up + 20 knots[2]	
"IAS." "A/C Override" "Set MCT" Call for the appropriate checklist After aircraft is properly trimmed, the Autopilot may be engaged, if desired	On command: – Select IAS – Select AIR COND SHUTOFF switch - OVRD – Select MCT on TRI/TRP and adjust thrust on operating engine to EPR called for – Accomplish the appropriate checklist on command

Traffic Advisory (TA) Actions & Callouts

WARNING

Comply with the RA vertical guidance and ATC lateral guidance if there is a conflict between the RA and air traffic control.

Once an RA has been issued, safe separation could be compromised if current vertical speed is changed, except as necessary to comply with the RA.

(OM Vol. 1, MANEUVERS 10.7.1)

Pilot Flying	Pilot Monitoring
TCAS traffic advisory (TA) occurs	
Look for traffic using traffic display as a guide	
Call out any conflicting traffic	
If traffic is sighted, maneuver if needed	

Note

If stick shaker or initial buffet occurs during the maneuver, immediately accomplish the APPROACH TO STALL RECOVERY procedure.

If high speed buffet occurs during the maneuver, relax pitch force as necessary to reduce buffet, but continue the maneuver.

Do not use flight director commands until clear of conflict.

Resolution Advisory (RA) Actions & Callouts - Not in Landing Config

WARNING

A **DESCEND** (fly down) RA issued below 1000 feet AGL should not be followed.

(OM Vol. 1, MANEUVERS 10.7.3

Pilot Flying	Pilot Monitoring
TCAS resolution advisory (RA) occurs	
If maneuvering is required, disengage the autopilot and autothrottle – Smoothly adjust pitch and thrust to satisfy the RA command – Follow the planned lateral flight path unless visual contact with the conflicting traffic requires other action	
Look for traffic using traffic display as a guide. Attempt to establish visual contact.	
Call out any conflicting traffic	

Climb RA in Landing Configuration Actions & Callouts

Note

When responding to an RA, the aircraft should be maneuvered only as much as needed to satisfy the RA.

If an RA response requires deviation from an ATC clearance, expeditiously return to the current ATC clearance when the traffic conflict is resolved, the "CLEAR OF CONFLICT" message is heard, or follow any subsequent change to clearance as advised by ATC. In responding to an RA that directs a deviation from assigned altitude, communicate with ATC as soon as practicable after responding to the RA. When the RA is cleared, the flightcrew should advise ATC that they are returning to their previously assigned clearance or should acknowledge any amended clearance issued.

Other critical warnings such as windshear or GPWS take precedence over an RA.

(OM Vol. 1, MANEUVERS 10.7.4)

Captain	First Officer
TCAS resolution advisory (RA) occurs	
– Disengage the autopilot and autothrottles – Advance thrust levers forward to ensure maximum thrust is attained	
"FLAPS 15."	
– Smoothly adjust pitch to satisfy the RA command – Follow the planned lateral flight path unless visual contact with the conflicting traffic requires other action	– Verify maximum thrust set. – Position flap lever to 15 detent
	Verify a positive rate of climb on VSI
	"Positive Rate"
"Gear Up."	
	Gear Handle .... UP
Attempt to establish visual contact.	
Call out any conflicting traffic.	

Moderate Turbulence Actions & Callouts

(OM Vol. 1, MANEUVERS 10.8.1)

Pilot Flying	Pilot Monitoring
Moderate turbulence is encountered	
Airspeed: – Above 10,000 MSL - 290 KIAS or mach .78 whichever is lower – Below 10,000 MSL - 250 KIAS or Clean Minimum Maneuver may be used whichever is greater Altitude: – Fly the FMS optimum altitude when possible to enhance buffet margin and economy – Descend if necessary to improve buffet margin	– SEATBELTS selector - ON – Make PA - See FM Part I - Section 11 - Non-Routine PAs Typical Examples - Turbulence – ATC - Notify • Alert ATC of any significant altitude deviation (300 feet or greater)

Severe Turbulence Actions & Callouts

[1] For FMS aircraft: in areas of turbulence, fly the FMS optimum altitude when possible. Buffet margin and economy will be enhanced. Descend if necessary to improve buffet margin.

[2] Below 10,000 feet, the greater of 250 KIAS or minimum maneuver may be used. Do not fly less than minimum maneuvering speed for existing configuration.

(OM Vol. 1, MANEUVERS 10.8.2)

Pilot Flying	Pilot Monitoring
Severe Turbulence is encountered	
– Autothrottle - OFF – ENG / AIR FOIL anti-ice switches - As required – Autopilot - Monitor • Use the autopilot in turbulence • Closely monitor autopilot operation and be prepared to disconnect the autopilot only if the aircraft does not maintain an acceptable attitude • If the autopilot disconnects, smoothly take control and stabilize the pitch attitude – ENG SYNC selector - OFF – Throttle - Set1 – Speed - Turbulent Air Penetration • 275 to 285 KIAS or MACH .75 to .79 whichever is slower	– SEAT BELTS switch - ON – ENG IGN switch - CONTIN or A or B – Make PA - See FM Part I - Section 11 - Non-Routine PAs Typical Examples - Turbulence – ATC - Alert • Alert ATC of any significant altitude deviation (300 feet or greater)

Upset Recovery
Nose High Recovery Actions & Callouts

WARNING

Excessive use of pitch trim or rudder can aggravate an upset, result in loss of control, or result in high structural loads.

(OM Vol. 1, MANEUVERS 10.9.2)

Pilot Flying	Pilot Monitoring
Nose high upset	
Recognize and confirm the situation	
– Disconnect autopilot and autothrottle – Apply as much as full nose down elevator – Apply appropriate nose down stabilizer trim – Thrust, as appropriate – Roll (adjust bank angle to as much as 60 degrees) to obtain a nose down pitch rate Complete the recovery: – When approaching the horizon, roll to wings level – Check airspeed and adjust thrust – Establish pitch attitude	Call out: – Attitude, airspeed and altitude throughout the recovery – Any omissions
	Verify all required actions have been completed

Upset Recovery
Nose Low Recovery Actions & Callouts

WARNING

Excessive use of pitch trim or rudder can aggravate an upset, result in loss of control, or result in high structural loads.

(OM Vol. 1, MANEUVERS 10.9.2)

Pilot Flying	Pilot Monitoring
Nose low upset	
Recognize and confirm the situation	
– Disconnect autopilot and autothrottle – Roll in shortest direction to wings level (unload and roll if bank angle is more than 90°) Recover to level flight: – Apply nose up elevator – Apply nose up trim, if required – Adjust thrust and drag as required	Call out: – Attitude, airspeed and altitude throughout the recovery – Any omissions
	Verify all required actions have been completed

Windshear Escape Actions & Callouts

WARNING

If ground contact is imminent, apply throttles full forward as needed to recover.'

CAUTION

Airline policy for low level windshear advisories is that takeoff is permitted, however caution should be exercised. Flights may not takeoff or conduct the final approach segment to a runway when ATC has reported a runway specific "Microburst Alert."

Note

Windshear escape maneuver is required during takeoff or landing where there is risk of ground contact.

Avoid over-boosting the engines unless necessary to avoid ground contact. When encountering windshear "Escape, set go-around thrust."

[1] If go-around thrust is exceeded at any point during the escape maneuver, make an AML entry.

[2] Do not exceed the pitch limit indication.

(OM Vol. 1, MANEUVERS 10.9.2)

Pilot Flying	Pilot Monitoring
When encountering windshear	
"Escape, set go-around thrust"	
Simultaneously: – Disconnect autopilot – Throttles - TOGA If ground contact is imminent: – Disconnect Autothrottle – Advance throttles full forward1 – Roll wings level and rotate toward an initial pitch attitude of 15° – Follow flight director commands (if available)2 – Turn flight director switches OFF, if WAGS is inoperative	– Ensure go-around thrust is set – Ensure all required actions are completed Call out: – Any omissions – Altitude and trend information based on radio altimeter (e.g., "300 feet descending." or "400 feet climbing" Retract speedbrakes
Do not: – Change gear/flap configuration – Attempt to regain lost airspeed until windshear is no longer a factor	
After escape is successful	
– Resume normal flight – Retract gear and flaps as required	Issue PIREP to ATC

Windshear Alerts: During Takeoff

[1] Inhibited from 80 knots to 400 feet RA.
[2] Inhibited from 100 knots to 50 feet RA.
[3] Inhibited until rotation.

(OM Volume 1, MANEUVERS 10.10.4)

Alert	Prior to V1	At or Above V1
Caution[1] "Monitor radar display."	Delay/reject the takeoff	– Maneuver as required to avoid windshear – Consider using maximum thrust
Warning "Windshear ahead. Windshear ahead."[2] or "Windshear. Windshear. Windshear."[3]	Delay/reject the takeoff	Perform the Windshear Escape Maneuver
Unacceptable Flight Deviations	Reject the takeoff	– Perform the Windshear Escape Maneuver – At VR, rotate normally to 15° no later than 2000 feet runway remaining

Windshear Alerts: During Approach

1Inhibited below 400 feet RA.
2Inhibited below 50 feet RA.

(OM Volume 1, MANEUVERS 10.10.6)

Alert/Aural	During Approach
Caution [1] "Monitor radar display."	Continue the approach if able to avoid windshear Otherwise, execute a normal go-around and maneuver as required to avoid the windshear
Warning [2] "Go around. Windshear ahead."	Perform either: – A normal go-around or – The Windshear Escape Maneuver
Warning "Windshear. Windshear. Windshear."	Perform the Windshear Escape Maneuver
Unacceptable Airspeed Deviations	Perform the Windshear Escape Maneuver

Normal Procedures Notes

Notes for this section focus on numbers and key items which are frequently needed in the cockpit, (and possibly on orals and simulators!) and for which there is usually not time to refer to the book. An example is for engine hot start considerations. In some cases, the information has been assembled from various places in the OM. In these cases, reference page numbers have been provided. Other then this type of item, no attempt has been made to summarize the NORMALS section.

DFGC/Autoland Test

NO AUTOLAND light flashes 50 seconds, then off
Headings must be ±2° between CA & F/O indicators

OM Volume 1, SYSTEMS 25.1

Standardized Actions

Callouts—Bold Italics Actions-Non Bold PM Actions GreenGear

Takeoff	2 Engine Go-Around	Takeoff Engine Failure	1 Engine Go-Around
Autothrottle—ON *Thrust Set* *80 Knots—Check* *V_1* *Rotate* *V_2* *V_2+10* *Positive Rate,* *Gear Up*	*Go-Around* *Set Go-Around Thrust* *Flaps – 15* (or *Flaps 11* as required) *Positive Rate,* *Gear Up* *Set and Arm Missed Approach Altitude*	*Autothrottle—ON"* *Thrust Set* *80 Knots—Check* *V_1* *My Aircraft* *Rotate* *V_2* *Positive Rate,* *Gear Up* *Runway Heading, HDG SEL*	*Go-Around* *Set Go-Around Thrust* *Flaps 11* **VREF + 5 Minimum** *Positive Rate,* *Gear Up* *Runway HDG, HDG SEL* *Set and Arm Missed Approach Altitude*
Passing 400' AGL Minimum	**Passing 400' AGL Minimum**		
HDG SEL or *NAV*	*HDG HOLD,* *NAV* or *HDG SEL* *Set Speed 200*		
	Passing 500' AGL Minimum		
Autopilot ON (if desired)			*Set Speed 250*
> 1000 AFL		**EOAA**	**> 1000 AFL**
Half Rate *Climb Power* *Flaps Up* (on schedule)	*Half Rate* *Climb Power* *Flaps Up* (on schedule) *Speed Select*	*Altitude Hold* If Engine Fire or Severe Damage: Memory Items Autothrottle...OFF Throttle Affected Engine..Confirm ....IDLE *Flaps Up* *Slats Retract* (On Schedule)	*Altitude Hold* *Flaps Up* (On Schedule)
0/EXT			**0/EXT**
Slats Retract			*Slats Retract*
		0/EXT + 20	
Clean Min Maneuver IAS, Bank 30 **2500' AFL** *IAS 250* or *SPD SEL 250* or *PERF* or *VNAV*		*IAS* *A/C Override* *SET MCT* *Checklist*	*IAS* *A/C Override* *SET MCT* *Checklist*

OM Vol. 1, General 15.4

Starting Engines

Normal Engine Idle Indications ..OM Vol. 1, Starting 15.6

- **APU** or **EXT PWR (L) POWER IN USE** Light ....Extinguished
- **L CSD OIL PRESS LOW** Light........................Extinguished
- **L OIL PRESS LOW** Light ....Extinguished
- **L HYD PRESS LOW** Light ....Extinguished
- EGT ....300-480°C
- Fuel Flow..600-1100 pph.
- Oil Pressure ....40-55 psi.
- N2 RPM...50 -61%

Engine Start Key Notes ....STARTING 15.7-15.8

Engine Warm-up—If the engines have been shut down more than two hours, warm up engines 5 minutes at low power settings.

Starting EGT

Tailwinds > 20 knots
Reverse engine rotation may occur
Increases probability of hot start
Closely monitor EGT and other parameters until engine is stabilized at idle RPM.

Normal starting fuel flow—approximately 600 pph
Starting fuel flow ≥ 1100 pph may be indicate a hot start
If observed, monitor EGT and other indications
Move the FUEL Lever to OFF when hot start is imminent
Observe starter duty cycle limitation while motoring

Starting EGT of 465°C through 500°C—
Make a numbered Info to Maintenance entry in AML
No Maintenance action is required.

If EGT limit 500° is exceeded for any period of time
Immediately shut down the engine
Record the peak EGT and duration of over temp in E6. Request maintenance to come out to the aircraft.

TWA For 94xx and 96xx airplanes
Engine data on electronic Engine Display Panel (EDP)
N_1, N_2, and EGT Indicator digits flash when a hot start is imminent or maximum value is exceeded

Idle RPM

Low Idle RPM
N2 RPM < 50% *accompanied by* EGT greater than 480 *and* possible generator cycling *may be an indication* 13th stage start bleed valve has failed to close
- Momentarily advance throttle to 65% N2, then retard to IDLE.
- Start bleed valve should close,
- Normal stabilized idle engine readings should result

If N2 < 50% or generator cycling persists, contact Maintenance
High altitude airports—May need to advance throttle > 65% N2
High Idle RPM—If N2 RPM > 67% contact Maintenance.

Post Start Overtemp

EGT is 480 – 590°C after stabilized idle
Engine shut down is not required.
Make AML entry

If EGT > 590°C,
Immediately shut down engine
Contact maintenance and make an AML entry including:

- Idle N2 RPM
- OAT
- Barometric pressure
- Fuel flow
- Peak EGT and duration
- Generator cycling, if any
- Throttle movement, if any
- Airplane movement, if any.

Engine Abnormal Start ....QRH, ENG 7.9

Use Engine Panel for cues
Top to bottom, (including Fuel Levers)
Each item has 1 associated item to remember ***except*** 2 each for N_2 and EGT

Fire Handles—Tailpipe fire or torching..........QRH FIRE 8.41
Oil Pressure Gage— ....QRH ENG 7.28
Pressure <40 psi or no rise ....QRH ENG 7.28, 31
OIL PRESS LOW Light on after start ....QRH ENG 7.33
Note—*For cold soaked engine (oil temp below 25°C), up to 5 minutes allowed for normal oil pressure.*
N_2 Gage-Air pressure to starter, no N2 rotation.QRH ENG 7.30
N_1 Gage—No N_1, confirmed by ground crew .....QRH ENG 7.29
EGT Gage—(EGT is Important, so **2** Items again)
No EGT rise within 20 seconds of fuel lever to **ON**
EGT > 500°C (ground) or 625°C (air).......LIMITATIONS 10.16

Items below have been removed from QRH and are included for reference only

Fuel Flow—Initial FF over 900 pph—anticipate hot start.......
Removed from QRH
Item not on panel—associate with fuel flow gage

Engine Start Switch—inadvertently released or positioned to **OFF** prior to 40% N_2
Allow N_2 to decrease to zero before re-engaging
If this occurs after fuel lever positioned to **ON**, immediately abort the start by placing the fuel lever to **OFF** ....Removed

Three Lights—signal need to abort after start sequence

- **CSD OIL PRESSURE LOW** Light remains on........QRH ELECT 6.28
- **HYD PRESS LOW** Light remains on ....OM Vol. 1, STARTING 15.6
- **START VALVE** Light—Start valve did not close.....QRH ENG-7.48

QRH Starting Procedures to Review

Engine Abnormal Start Procedure ....QRH ENGINES 7.9
Engine Clearing ....QRH 7.11

Other Power Limitations...........OM Vol. 1, TAXI-TAKEOFF 10.3

One-engine taxi—avoid over 1.2 EPR for breakaway power in vicinity of gate area

To start immediate turn
- Set thrust and wait for the airplane to respond
- Roll forward before turning

Before Takeoff

Flaps/Slats Callout

11/ ....Closeout Flap Setting
11/ ....Flap Thumb wheel
11/..Flap Handle Position
11/ ....Flap Position Indicator
TAKEOFF.................................Slat "TAKEOFF" Light ON

Takeoff

Takeoff Bugs

WHITE V1
ORANGE V2
WHITE ....Flap Retract
WHITE ....Slat Retract
WHITE ....Clean Minimum Maneuvering
RED COMMAND ....250 Knots

Final Power Adjustment

Complete by 60 knots
Autothrottle CLMP by 60 knots or AT must be disengaged

Takeoff Attitude

8° Desired
10.5° = Tail Strike

Climb Speed Safety Margin

V_2 + 10 yields 20% margin over stall (1.2 x V_{STALL}) for bank angles up to 15°

If 30° bank is required, use V_2 + 20, which yields a 30% stall margin in a 30° bank turn

Special airport procedures listed in Part II are designed to ensure a 20% over V_{STALL} margin

After Take-Off Climb

Initial Climb Speeds

250 Knots or (clean MIN MAN, whichever is greater) until 10,000 feet MSL
120,000 Lbs or less—290 Knots until planned cruise mach
Over 120,000 Lbs—300 Knots until planned cruise mach
Cruise Mach above Mach crossover to cruise altitude

Descent

Standard Descent Speeds for Planning

Arriving at domestic US stations, plan to descend using the following descent schedule:
- ♦Cruise Mach to the 290 knot crossover altitude
- ♦290 knots to 10,000 feet
- ♦250 knots or less below 10,000 feet

Outside domestic airspace, consider an optimum (CI generated) descent speed into those destinations where experience and judgment indicate no ATC conflict.

Airspeed Bug Settings

WHITE ....0/Ret—Min Maneuver
WHITE ....0/Ext—Min Maneuver
WHITE ....15/Ext—Min Maneuver
....11/Ext if required by Airport Analysis
COMMAND BUG ....Approach Speed
- For all conditions unless specified in abnormal procedures.....
  - Between V_{REF}+5 and V_{REF}+20
- A/T OFF—V_{REF} + greater of: ....5 Kts (VREF + 5)
  - ....*or* 1/2 steady state wind above 20 Kt
  - *or* all of the gusts above steady winds
- A/T ON...........ATS speed 5 knots less than computed speed
  - ....with corrections above, not less than V_{REF}+5
- Autolanding ....V_{REF}+5 with no wind additives
  - See OM Vol. 1, CL-CRUISE-DES-20.5

ORANGE ....VREF
WHITE ....80 Knots

Before Landing

Flaps & Slats Callout

40/ ....Flap Handle Setting
40/ F ....Flap Position Indicator

Landing

Landing Flap Selection APP-LDG-G/A-10.2-10.3

Flaps 28° and 40° are normal landing settings

Flaps 40° *Required* for anti-skid INOP
Recommended when:
- ♦Runway ≤7000
- ♦Runway wet/slippery
- ♦Braking action less than good
- ♦Tailwind, crosswinds, or gusty wind (*velocity not specified*)
- ♦A/C weight very light (not defined)
- ♦Steeper than normal approach due to ATC requirements

Auto Brake Landing Settings ....APP-LDG-G/A 15.3

Required when:
- ♦Runway <7000'
- ♦RVR <4000' or visibility <¾ mile
- ♦Runway contaminated (standing water, snow, slush or ice)
- ♦Braking action reported less than good

Recommended when:
- ♦Landing with "gusty winds or crosswinds" or tailwinds

Setting selection: "Autobrake settings should be appropriate to the conditions: **MAX** must be used when minimum stopping distance is required."
OM Vol. 1, APP-LDG-G/A-15.3, 30.5, 40.3

Thrust Reverser Considerations ....APP-LDG-G/A 45.3

With spoilers deployed *and* directional control assured reverse thrust <u>may be left at idle</u> *or* <u>increased</u> to a target of approximately 1.3 EPR

Application of reverse thrust tends to blank out the rudder
- Rudder effectiveness starts decreasing with the application of reverse thrust and at 90 knots
- A 1.6 EPR (in reverse) rudder is almost completely ineffective

Reverse thrust > 1.3 EPR should not be used unless stopping distance is in doubt

Do not exceed 1.3 EPR reverse thrust on wet or contaminated runway, except in emergency, when max reverse thrust may be used

If the airplane starts drifting across runway while reversing,
- Immediately return thrust levers to idle reverse to regain directional control and restore rudder effectiveness.
- Do not use asymmetrical reverse thrust to regain directional control
- Reapply reverse thrust when directional control is restored.

When using more than idle reverse thrust, rapid movement of the reverse levers to the stowed position will cause the airplane to accelerate as the engines are decelerating to idle RPM.

CAUTION

Moving the Reverse Levers to reverse idle prior to the nose gear being firmly on the runway may cause the reverse buckets to contact the ground.

The combination of forward airplane movement and reverse thrust at airspeeds below 60 knots results in an airflow capable of causing ingestion of runway / taxiway debris into the engines. This condition should be avoided whenever possible.

Manual Braking....................APP-LDG-G/A -45.4, 50.11

Whenever practicable, do not use brakes above 100 knots

If stopping distance is critical, or other abnormal conditions exist, brakes may be used from touchdown to full stop.

See additional notes, page APP-LDG-G/A 50.15

Approach Techniques

Initial Setup..**WARM**

Weather requirements ....Check
Include crosswinds, visibility, published minimums

Approach.................................Plan and Brief Approach

RNP..........Check Chart and set as required for the approach

MCP—When cleared for the approach ....Select APP Mode
....Set FAF altitude when on intercept ***or*** a portion of the approach

Captain Callouts................LOC / GS Intercept / Passing FAF

Passing FAF Actions..**MAC-V**

Missed approach Altitude ....Verify Set

APPR Light—Green (RNAV Approaches) ....Verify

Clearance—Tower ....Call Passing FAF

V/S..Set as required

Note: *This section compiled from instructor notes and various OM references; NOT procedural, but technique recommendations.*

Approach Deviation Callouts

Airspeed error exceeds -5 or +10 knots

Rate of Descent

When Below:	If Descent Rate Exceeds:
2000' AFL	2000 fpm
1000' AFL	1000 fpm
Inside FAF	1000 fpm

Localizer and Glide Slope

Localizer— 1/3 Dot on PFD localizer indication.........*COURSE*

Glide Slope—1/2 Dot on glide slope..................*GLIDESLOPE*

RNAV GPS / GNSS

Course—1 dot deviation ....*COURSE*

Vertical deviation < 1,000' AFL— 1/2 Dot on VDI....*GLIDESLOPE*

Other Approaches & **Parameters** ....OM Vol. 1, APP-LDG-G/A 20.4

Briefings and Restrictions

Flight Attendant Briefing—Prior to gate departure

Number of F/A's on board

Known delays

Ground—Short-taxi & safety demo considerations

Turbulence

- Weather and turbulence forecast for route
- Unexpected turbulence
- Notification
- Required action in the event of a significant encounter
  - Service suspension at F/A's discretion
  - When able, call CA to report cabin situation / injuries
  - No access to interphone—stay seated, avoid injury
  - Resuming Duties
  - Once cleared to resume call flight deck to report any injuries or abnormalities.

Security items

E6 Cabin Items

Life vest demo—overwater segment

Gen Decs/Customs Immigration forms (if required).

Cockpit Access—Crew meals / Cockpit Door

Cabin Door Handles—Push Down

Miscellaneous—Issues relevant to flight ....FM Part 1, page 7.1-3

Restricted Captain Requirements ....FM Part 1, 4.2-1

Applies if

CA has < 100 hours after IOE in current aircraft type

Required 100 hours PIC in aircraft type may be reduced not to exceed 50% by substituting one landing for one hour of PIC time in aircraft type.

Restrictions *unless* FAA Exemption 5549 apply

- ♦ Chart increases required visibility ....FM Part 1, 4.2-1
- ♦ Alternate Airport - Published minima need not be increased, but lowest allowable landing minima is 300 feet MDA/DH and one mile visibility or 4500 RVR / 1400 meters

Practical Effect of Exemption 5549:

Recognizes inherent safety of autopilot coupled approaches

Allows approaches using

- ♦CAT II procedures to no lower CAT published minima or
- ♦CAT I procedures to published CAT I MDA, no lower than RVR 1800'

See detailed charts ....FM Part 1, pp. 4.2-1 to 4.2-3

Low Experience FO: <100 Hours (CFR 121.438)

When Applies—FO has <100 hours flight time as SIC in type aircraft being flown, ***and***
Captain (PIC) is not a qualified Check Airman,

PIC must make all takeoffs when:

- All special requirement airports
- Visibility in the latest WX report at or below 3/4 mile
- RVR for the runway to be used is at or below 4000 feet
- Runway has water, snow, slush, or similar conditions which may adversely affect A/C performance
- Braking action is reported less than good
- Crosswind component >15 kts
- Windshear reported in the vicinity of the airport
- Any other condition PIC determines prudent....FM Pt. 1, 4.2-4

Flight Attendants—Emergency TEST Acronym

T—TYPE of emergency: General description of the emergency

E—EVACUATION: Yes or No
Include any special considerations which may affect the use of exits.

S—SIGNAL: the following PA:
"This is the Captain. Evacuate. Evacuate. Evacuate."
Followed by turning on Evacuation Command (as installed).

T—TIME to landing.........................QRH MISCELLANEOUS 0.11
....FM Part 1, 19.2.3

Emergency PA to Passengers

Nature of emergency

Time to landing

Passenger cooperation with Flight Attendants

Reassure passengers. ....QRH MISCELLANEOUS 0.12

Cold Weather Operations

Cold Weather Operations and De / Anti-icing Guidance
.......OM Vol. 1, GENERAL 30.1 to 30.22 and 31.1 to 31d-5

Icing Definition—Ground or Flight

Ground or Flight RAT—Below 6°C / 42°F

***AND* Visible moisture present** such as clouds, fog with visibility one mile or less, rain, snow, sleet, or ice crystals.

OR (Ground only) Temperature as above ***and*** temperature-dew point spread is 3°C (5°F) or less

AND Any form of moisture is present (standing water, snow, slush etc.) ***which may be***
Ingested by the engines and freeze on engine inlets, nacelles or engine sensor probes

Preflight

Advise Dispatch if RLW needs adjustment

Special Attention **for ice in:**

- ♦Wheel wells, actuators & steering components
- ♦Flight controls (flaps, slats, control tabs, etc.)
- ♦Wing surfaces (Max 1/8 inch frost on bottom
- ♦Pitot-Static components
- ♦Engine Inlets
- ♦Pack inlet/exit doors
- ♦Fuel tank vents
- ♦Pressure reg./relief valves
- ♦APU Inlet

Check area behind overwing heater for ice runback and lower wing surfaces for icicles.

Upper wing surface inspection required by cockpit crewmember when in Cold Weather temperature range

Operative Overwing Heater System does not relieve flight crew of requirement to check airplane surfaces free of frost, snow, and ice accumulation, as required by FARs

Physical feel (pole) check required if heater blankets inoperative and the temperature is less than 10°C (50°F) and high humidity or visible moisture is present.

Frost on the bottom of the wings under the fuel tanks may occur when the fuel temperature is low, OAT is above freezing and humidity is high

Deicing

Packs and Bleeds (Engine & APU) OFF During and for 1 minute after deicing
APU—May be ON
Engines If operating, at idle power
Before T/O Checklist Must be completed in entirety following deicing

Start/Taxi

Caution: *If frozen contamination in intakes, must be removed by certified deicing personnel.*
Engine Start—Ensure normal N_1 during start
Oil Pressure may be high due to low OAT
- ♦Must be in normal range when oil temperature stabilizes in normal range
- ♦Operate at idle until oil pressure reaches normal range

Engine Anti-Ice ON Immediately after start if icing conditions exist or anticipated before T/O
Ignition **CONTIN A** or **B**
Flaps & Slats Up until just before takeoff ***if***
Taxiway contaminated *or*
Freezing precipitation continues after deice/anti-ice
Caution—*Monitor flap indicator for early stop*
Flight Controls—Check for full travel of flight controls
Periodic engine run-up recommended no more frequently than every 10 minutes
70% N_1 for 15 seconds or 60% N_1 for 40 seconds
Use caution for *jet blast*
Any Precipitation after holdover time—visual inspection required *(Cockpit observation of wipers, etc., —use to determine need for ext. insp.)*
Fuel Heat before takeoff if fuel <0°C
One minute each
First engine stabilized before starting heat on second engine at a time
Fuel heat off for takeoff ....OM Vol. 1, SYSTEMS 65.1

Contaminated Runways

A runway should be considered contaminated when:
More than 25 percent of the required field length, within the width being used, is covered by:
- ♦Standing water, slush, or wet snow deeper than 1/8“ (3 mm)
- ♦Dry snow deeper than 1 inch (25 mm),
- ♦Ice.

If a runway is contaminated:
- ♦Takeoff is not authorized with a tailwind
- ♦Takeoff is not authorized with
  - ♦♦More than 1/2 inch of wet snow, slush, ***OR***
  - ♦♦Standing water, ***OR***
  - ♦♦More than 4 inches of dry snow
- ♦Maximum thrust must be used
  If ART (MEL item 73-8) is inoperative use Reserve thrust. Standard thrust is not authorized
- ♦Both thrust reversers must be operative
- ♦APU will be used for takeoff, if operative.
  - ♦♦APU Air Switch is **OFF**
  - ♦♦APU Bus Switches are **ON**)
- ♦Takeoff not authorized with chunks of hardened snow or ice
- ♦Corrections to V1 and maximum weight allowances are made by dispatchers and sent to aircraft

Performance, TAKEOFF 40.1
See also page 54 of this Study Guide

Takeoff

Runway Is unfit for takeoff if:
>4" Fresh, dry snow
½" Standing water/wet snow/slush
Chunks of ice or hardened snow ....PERF TAKEOFF 40.1
Run Engines Up to 1.4 EPR or 80% N1 and check for normal indications prior to brake release
Align A/C with runway before T/O Power is set
Cross-Check Engines instruments for “reasonableness”
Asymmetric Thrust—Avoid (directional control)
Rotation—Slow & smooth—avoid abrupt or early
400' or Above—Packs ON IAW CL-CR-DESC p. 10.4
1000' or Above—Wing Anti-Ice ON ....Performance, TAKEOFF 40.1

Climb, Cruise, Descent

Sufficient Thrust to keep L or **R ICE PROTECT TEMP LOW** Light out
Watch for Drop in thrust parameters (Icing of PT2 probe indications)
Airfoil Anti-Ice min duct pressure 20psi
Tail Deicing cycled every 20 minutes
Fuel Heat and tail deice cycle 1 minute prior to approach
Severe Icing conditions—
Ignition OVRD
Minimum N1—70%
If N1 reduction is necessary, no lower than 55%
After reduction, N1 to 75% when able for 1 minute (minimum)

Landing

"In the Slot" extremely critical with poor braking
- ♦≈250' at ¾ Mile Final
- ♦On Glideslope
- ♦Sink Rate Appropriate
- ♦Trimmed Up
- ♦Aligned with Runway
- ♦Thrust Steady
- ♦Final Landing Configuration

40° Landing recommended on wet or slippery runways
If a landing on a runway contaminated by standing water, slush, snow or ice:
- ♦APU (if operative) will be started and
- ♦Left and right APU bus switches ON prior to final approach.
  Serves as backup electrical power source in case the engine driven generators are lost due to slush or water ingestion by engines and subsequent loss of engine RPM

Recommended techniques:
- ♦Land on speed.
- ♦Touchdown at the planned point
- ♦A firm landing is better than a “grease job.”
- ♦Keep nose wheel firmly on runway with elevator.
  __Caution__: *Excessive down elevator force will download the main gear and reduce braking efficiency.*
- ♦Use maximum auto brakes or aggressive manual braking and auto spoilers, if available
- ♦Maintaining directional control is the highest priority
- ♦Apply reverse thrust
  As soon as possible after nosewheel touchdown.
  Do not exceed 1.3 EPR reverse thrust on the slippery portions of the runway, except in an emergency
- ♦When reversing, if directional control is lost:
  Reduce reverse until control is regained
  Use forward idle thrust if necessary
- ♦Do not come out of reverse at a high RPM
  Sudden transition of reversers before engines spool down will cause forward acceleration.
- ♦Use as much of the runway for roll-out as needed to slow airplane to a safe taxi speed before turning off a wet / slippery runway.
  __CAUTION__: *In an emergency, use maximum reverse thrust, if required, to stop in the remaining runway.* Runway Unfit for landing:
  1' of Standing water, wet snow, or slush
  Chunks of ice or hardened snow

Plan Firm Touchdown
Use Auto brakes/Auto spoilers
Apply Reverse Thrust ASAP after nosewheel touchdown:
Don’t exceed 1.3 EPR on slippery portions of the runway except in emergency.
In emergency, use maximum reverse thrust if required to stop in remaining runway

Taxi In & Parking

If Slush Or Snow on runway or after making an approach in icing conditions:
Leave flaps 15°
Call station for inspection by maintenance
Draining Water System required if
A/C Parked overnight in freezing temperatures
Lavatory fluid must be drained or treated anti-freeze
Call Dispatch to report runway conditions if appropriate
Call Station to coordinate inspection of flap area

References

OM Vol. 1, GENERAL 30.1 - 30.38,
OM Vol. 1, LIMITATIONS 10.28-10.30

Airplane General

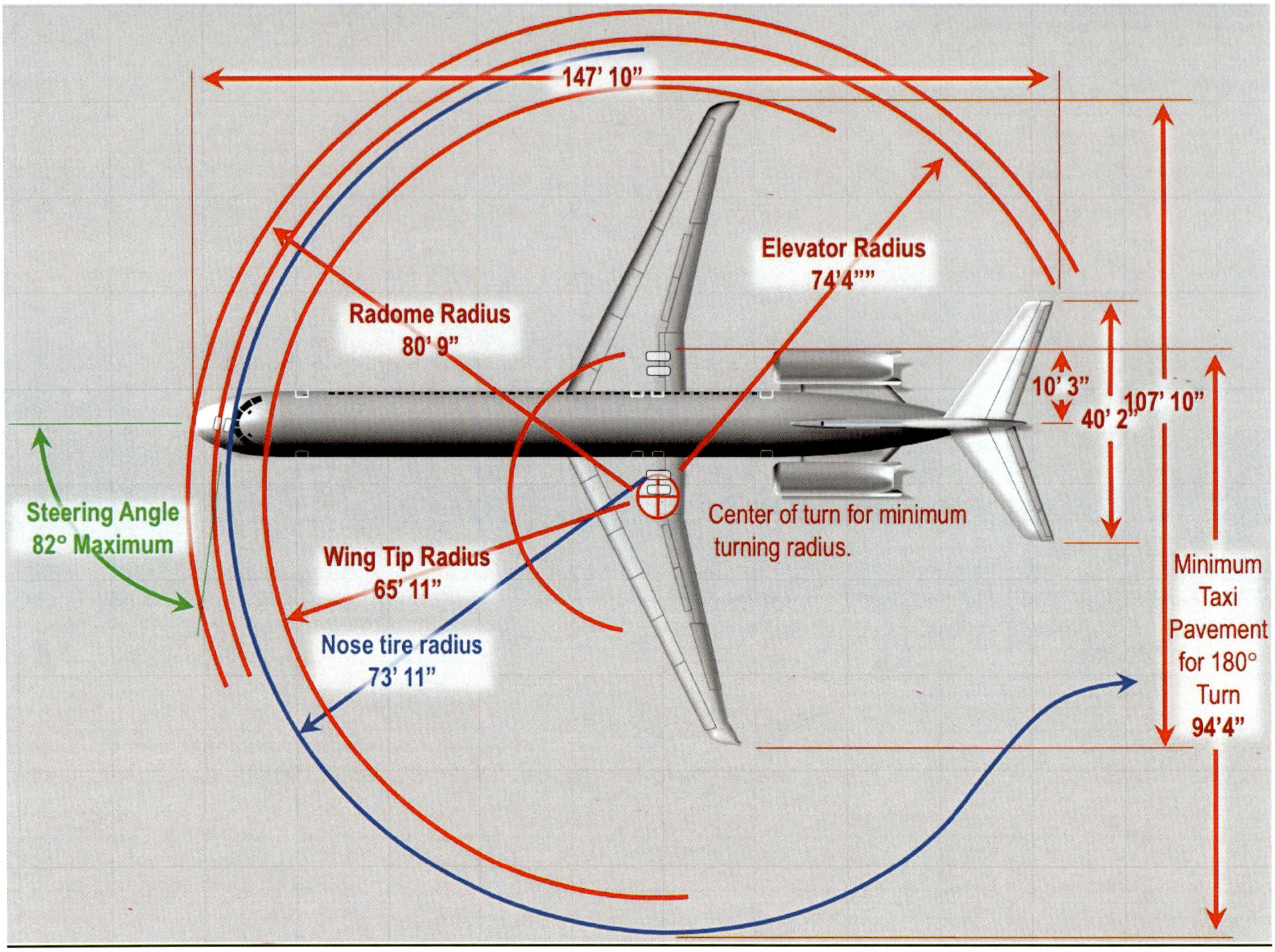

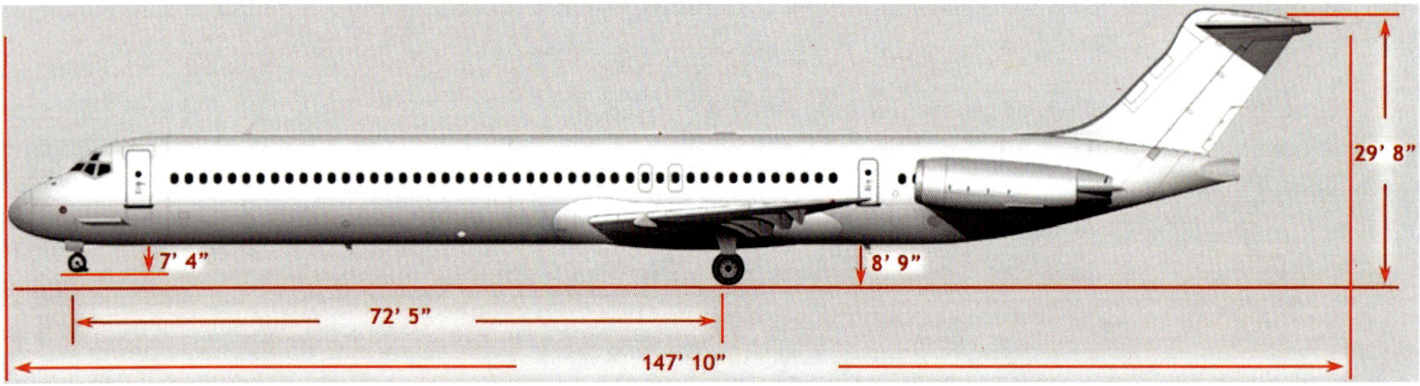

Side Silhouette drawing courtesy of ***Norebbo Stock Illustration and Design***, www.norebbo.com.

Air Conditioning and Pressurization

General

System Components

2 Air conditioning-pressurization units (packs)
- One on each side of the aircraft
- Normally, air supply from respective engine

Pneumatic Crossfeed Valves allow:
- Air from one side to run the pack on the other side
- The APU to supply air for either or both packs
- The engine to supply air to the airfoil anti-ice system

2 Independent/redundant pressurization controllers alternate roles as primary/backup each flight

Manual outflow valve control backs up electronic controllers

Positive relief valves prevent over-pressurization

Negative pressure relief through
- Inward movement of galley & passenger door seals
- Negative pressure relief valve on aft pressure bulkhead

AC—Pressure Regulated, Temperature Protected

AI—Temperature Regulated, Pressure Protected

Pack Automatic Shutdown Circuit

Shuts off both packs if engine fails on takeoff

<u>Arming</u>—after engine start when: (EPAD)
- *Engines—one or both running*
- *Pack Supply Switches (1 or both) HP BLD OFF or AUTO*
- *AUTO selected—Air Conditioning Shutoff Switch*
- *Differential Pressure (cabin to ambient) < 1.3 psi*

Activation:
- Differential 13th stage pressure (delta P) of 70 psi or more as failed engine spools down
- False activation can occur on single engine taxi with high power

<u>Deactivation</u>: Remains armed on ground and airborne through approx. 3000 feet AFL (1.3 psid point reached)

<u>Reset</u>—Air Conditioner Shutoff Switch to OVRD

System Inop—Make No Pack Takeoff

Pack Control

L and R Supply Switches AUTO
- Opens Flow Control and Pressure Regulator Valves to let air into the packs
- (Ground) Turns on Pack cooling fans for respective side

HP BLD OFF Position:
- <u>Opens:</u> ♦Pressure Regulator Valve ♦Flow Control Valve
- <u>Closes:</u> ♦Augmentation Valve
- ♦*Ground only*—Heat Exchanger Cooling Fan **ON**
- Keeps 13th Stage air out of system by maintaining augmentation valve closed
- Starts heat exchanger cooling fans

Desired output pressure—approx. 21 psi
- With engines running—13th stage augmentation valves modulate as necessary to augment 8th stage air
- With APU running—APU load control valve regulates air pressure output

Packs

Pack cooling fans
- Draw air across primary heat exchanger on ground with supply switches ON
- Cooling air diverter valve closes when fan is on

Overheat protection
- Thermal shutdown switches sense excessive temperatures in:
  - Compressor discharge
  - Turbine inlet
  - Outlet duct
- Result—Air conditioning flow control valve closes, effectively tripping the pack
- Reinstatement is automatic when temperatures return to normal
- No pilot action is necessary or available

Low-flow protection
- Pack may shut down if lower than 12 psi output sensed
- With flow gages—9 O'clock position

Air Sources:

Ground Pack Air Sources

♦Engines ♦APU ♦External high-pressure air

Other conditioned air source: External pre-conditioned air

Airborne Pack Air Sources

Engines
- Either can supply sufficient air for both packs
- Pneumatic crossfeeds allow airflow to opposite side

APU—*not* available
- In-flight use is for electrics only
- No airborne bleeding of APU air is allowed

Air Conditioning

Dual Temperature Controls

Manual and Automatic ranges

Normal air distribution
- Right pack supplies air to cabin
- Left pack controls air to cockpit and cabin

Recirculating Fan

Recirculates Cabin Air

Increases flow rate and reduces hot/cold pockets

Cockpit control:

Recirculation Fan Switch
- **OFF**—Self Explanatory
- **ON**—Allows recirculating fan to operate on the ground to supplement air conditioning
- **AUTO**—Off on ground, On in flight

APU Ground Cooling Use

All aircraft now modified with -280 APU and should provide sufficient airflow for cooling on ground

APU Air Switch AIR COND COLDER Position
- ♦Closes turbine bypass valve
- ♦Increases differential pressure across air conditioning turbine, lowering temperature of conditioned air

When using APU to supplement engine air, do NOT select **AIR COND COLDER** Position ....Vol. 1, PRE-FLIGHT 10.7

(**ON** Position provides best combination of air flow and refrigeration for this condition)

When using APU for electrical power, with two A/C units operating may exceed 1.25 AC Loadmeter limit

See Volume 1, SYSTEMS-15.3

Manual Temperature Control

Manual range on auto-manual temperature control rheostat

Max gage temperature is 150°F
- Do not exceed 150°F in supply duct
- Pack output >190°F (no indication) trips pack

Operation in MANUAL mode—
- <u>*Not recommended*</u> unless the automatic system has failed

<u>**Note**</u>: **CKPT TEMP Valve movement** is inhibited toward hot if left pack discharge temperature exceeds 130°F. Resets when temperature drops back below 130°F.

Pressurization

Key Pressures & Altitudes

1.3 psid—Over 1.3 psid prevents arming of the Pack Automatic Shutdown Circuit

7.5 psid—Manual pressurization maximum stable pressure

7.77±.3 psid—Normally programmed maximum pressure

7.77 psid— TWA Maximum Cabin Differential Pressure

8.07 psid—Maximum Cabin Differential Pressure

8.32 psid—Maximum Emergency Pressure Relief

10,000 Feet Cabin Altitude—CABIN ALT Warning light, Modulating horn for 1 second, followed by *"CABIN ALTITUDE"* aural warning (May come on ≈9,500')

Fasten Seat Belt signs illuminate automatically when **CABIN ALT** Warning light ON

10,000' MSL—Maximum Dispatched altitude for Unpressurized Flight

14,000' MSL—Maximum altitude following an in-flight depressurization **14,000' Cabin Altitude (Approximate)** —Passenger Masks deploy

25,000' Cruise Altitude—Max for dispatch, single pack

Cabin Climb & Descent Rates

With rate knob on index:
- Approx. 700 fpm climb
- Approx. 300 fpm descent

Range of control
- Climb—100-2000 fpm
- Descent—40-860 fpm

Pressure Controllers

Automatically trade legs—Changeover occurs at each landing

Designated "Primary" and **"Standby"** respectively

Powered by separate AC buses

Standby continually monitors pressurization as controlled by the primary pressure controller

If a primary controller discrepancy occurs:

Standby system takes over

STDBY ON light comes on

TRANSFER LOCKOUT light comes on

Reset should not be attempted if **STDBY ON** light is also ON

If **STDBY ON** is <u>***not***</u> illuminated, <u>***and***</u> there is no fault in the standby system, <u>***then***</u> pressing **TRANSFER LOCKOUT** light will reset the automatic transfer capability

If both lights come on in flight—

Cycle system to STDBY and back to PRIMARY

If STDBY ON light goes out, fault has cleared and control returned to primary controller

May have been caused by explainable anomaly such as:
- ♦Power fluctuation
- ♦Situationally-driven pressurization abnormality such as an unusually high rate of descent

Primary system may be used; reset **TRANSFER LOCKOUT** light

If **STDBY ON** light stays on
- Primary system has failed
- No further crew action required
- No reset of TRANSFER LOCKOUT should be attempted in flight to allow maintenance to troubleshoot

If TRANSFER LOCKOUT Light ON by itself

Discrepancy sensed in the standby controller-

Corrective action:
- Attempt Reset
- If light stays on—STDBY system has failed
- Further resets should not be attempted

Cabin Pressure System Selector Switch

Selecting either 1 or 2 makes that the primary system

Remaining system is standby

Standby system fully monitors itself

If selected system fails
- Blue **INOP** light illuminates
- Cabin pressure control transfer to remaining system is automatic

Auto Switching
- Occurs after landing
- Primary relinquishes control to standby for 30 second test
- Selected system then takes control back

If both systems fail—AUTO INOP light (*Amber*) .... TWA

<u>**Note:**</u> *Discussion of abnormals is for systems description only. Refer to QRH, AIR-5 for specific procedure.*

Both the TRANSFER LOCKOUT & STDBY ON lights are <u>*inoperative*</u> with the Cabin Altitude Control Lever in the **MANUAL** (Down) position

Cabin Pressurization Initiation

Occurs at takeoff power on the ground

If throttles are retarded or **no takeoff 60 seconds after power up**—depressurization occurs

System re-pressurizes with no crew input on next power up occurrence

Displays on Cabin Climb Gage

Outflow Valves

Two moveable exhaust doors
- Cabin Air Outflow Butterfly Valve—Controls pressurization at low differential pressure
- Cabin Air Outflow Nozzle—Used at high velocity to reduce drag

Manual control
- Both outflow valves de-clutched from actuator
- Valves are mechanically interconnected to sequence properly

Cargo Compartment Pressurization & Temperature Control

Pressurization—All compartments pressurized by intermittent action of cargo compartment pressurization equalization valves

Temperature
- Forward—May be heated by radio rack exhaust air ducted under lower cargo liner
- Mid—Heated by passenger compartment exhaust air
  - Ducted under lower cargo liner
  - Continuously circulated by a fan
- Aft—No temperature control

LDG ALT Control

Set field elevation for departure or landing altitude as appropriate

NOT like some controllers, where destination elevation is set prior to departure

Warning Lights & Other Systems

Flow Light

Non-TWA Aircraft—Flow light not used
- A placard-**FOR MAINTENANCE USE ONLY** is installed over the Flow light

Indicates cabin climb with fully closed outflow valve
- Likely causes
  - Insufficient air conditioning *OR*

Excessive fuselage leakage

Should be monitored in normal system scans .... **TWA**

Pressurization Flow Light MC

Light on in flight
- Cabin is climbing with a fully closed outflow valve
- Remains on until cabin pressure air flow is attained

Light is tested during the normal Annunciator/Digital Light test on preflight

Has replaced old **FLOW** light on all airplanes
- Light illuminates when **LTS TEST** Button is selected.

Illuminates when tested, but deactivated .... **TWA**

Tail Compartment Temp High Light MW

Tail compartment temperature is higher then normal

Emergency procedure eliminates all sources of 13th stage bleed from the system

Air Conditioning SUPPLY TEMP HIGH Lights MC

Lights come on when L or R air conditioning supply temp is over normal operating temperature. Indicates malfunction of the augmentation valve

Ram Air Switch

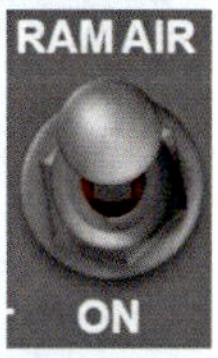

ON—Opens ram air valve
- Admits ram air to conditioned air ducts in right A/C system at a point upstream of the mixing chamber

OFF—Closes Ram Air Valve

Radio Rack Fan Switch

Inoperative on ground
- Primary and standby fans **ON** regardless of switch position
- Venturi closed

Flight
- **FAN**—
  - Primary fan ON,
  - Standby fan OFF unless primary fan fails, then ON
  - Venturi closed
  - Cools radio rack
  - Heats forward cargo compartment for animals and plants
- **VENTURI**—
  - Both fans OFF
  - Venturi open
  - Cools radio rack
  - Does not provide heat to forward cargo compartment

Supplemental heater
- Only primary fan has one
- Standby fan does not have a heater and will not heat cargo compartment

RADIO FAN OFF Light

Ground—Primary fan is inoperative

In Flight—
- Radio rack switch is in fan
- Both primary and standby fans are inoperative

Cabin Altitude Light (Red) MW

Comes on when cabin pressure exceeds 10,000' pressure alt.

When light comes on:
- ◆ .Warning horn & aural voice warning alternately sound for 5 seconds (1 second each, alternating)
  Aural Warning .... CABIN ALTITUDE
- ◆ .Cycle continues for 5 seconds or until condition is corrected, whichever occurs first
- ◆ .Fasten Seat Belts signs come on in cabin, as well.

Instrument Airflow Indicator (Ground operation only)

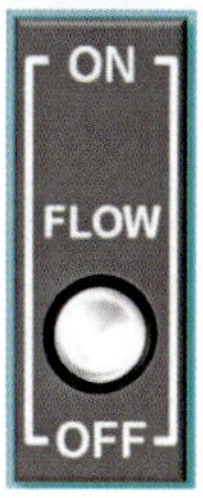

ON—Indicates flow is sensed to instrument panels

OFF—Indicates failure of instrument cooling fan

Above Gear position indicator lights on F/O panel

Note: Indications significant only on ground with normal electrical power applied and air conditioning **OFF**.

APU—Auxiliary Power Unit

General—Source for:

Pneumatics	Electrical
Air Conditioning	Ground—normal aircraft functions
Engine Start	Airborne—alternate electrical power

Fuel

Primary source—right main tank

Boost pumps which supply fuel to the APU

- DC Start pump—"Short Life" pump
  - Used when only battery power available
  - Normally only used for starting
- Any right main or center tank boost pump
- L main tank boost pump with crossfeed valve ON
- Normally used—RH Aft Fuel Boost Pump

Bleed Air System

Bleed air source—port on APU turbine plenum housing

Bleed airflow—selected On-Off in cockpit

One-way valve prevents reverse flow from engines to APU

Electronic Control Unit (ECU)

- Controls bleed pressure without loads to 15-22 psi
- With a pack or engine start valve energized open, pressure increases "to its normal operating range for this condition within 6 seconds"

Electrical vs. Bleed Air Load

Electrical load has higher priority than bleed air

With excessive EGT, load control valve restricts bleed air

APU Inlet Doors

1 Ram Air Door—allows ram air to assist APU RPM acceleration for airborne starts

2 Smaller non-ram air doors

May be sequenced automatically or manually

Door control sequence:

- APU Master Switch to START
  - Ram air door opens; Power is routed to APU Starter
- 35% RPM—ECU switch removes power from starter
- 95% RPM—Ram air door closes as non-ram doors open
  - Doors remain in this condition until APU shutdown

Door Control—DC Transfer Bus

Automatic Warm-Up/Cool Down Timer

Prevents bleed air load until 60 seconds after 95% RPM

APU Master Switch OFF—60 sec. APU shut down delay

Fire control switch OFF & AGENT ARM bypasses 60-second timer

APU Automatic Shutdowns (FOIL'D)

Warning Lights: All light the Master Caution/Warning Light **MC** **MW**

♦**APU Oil Press Low** light **MC**
  Self explanatory

♦**APU Fire**—APU Fire Detection System is activated. (See OM Vol. 2, FIRE PROTECTION-5) **MC** **MW**

♦**APU Fault**—Simply says a fault has been detected **MC**

Ground—Automatic shutdown, or no start

Airborne—APU Runs until ≈ 10 minutes after landing unless these faults are sensed ***FOIL'D***

Fire (APU)	**D**C Power loss
Overspeed	
***I**nternal Fault*	**For any of these five,**
***L**oss of speed sensing* **airborne**	**shutdown occurs**

APU Start Notes

Electric start capabilities up to 37,000 feet

> **Max electric start** capabilities up to 30,000 feet
> **Max operating altitude** 35,000 feet .... **TWA**

OM Volume 1, Systems 20.2

Air starts

- Ram-air-assisted electrical start
- No electric power—starter locks out; ram air only

Both AC Generator Buses unpowered—electric starter locks out for windmill-only start

Windmill Starts—No RPM indications may occur for over 1 minute; complete cycle may take in excess of 2 minutes

Battery required ON for start & APU running—APU control CB-battery bus

Should check bus control circuits OFF before start

RAM AIR and air conditioning SUPPLY switches must be OFF prior to start ....OM Vol. 1 PRE-FLIGHT 10.5

Momentary EGT/RPM gage deflection on start is normal

May take up to 55 seconds after first RPM indication to reach operating RPM (Don't mistake for hung start)

Starter Duty Cycle begins at first indication of APU RPM

Put MSTR Switch in RUN, wait 1 minute, then **START**

- RPM Indication should occur immediately
- APU Low Oil Pressure light on (out by 35% RPM normally, must be out by 95%—see below)

No Start or Hung Start

- Indications of Hung Start:
  - RPM Stabilized below normal range (95-105%)
  - EGT rising or near maximum
- Shut down with **APU MASTER** Switch ....QRH APU-2 & APU-3

By about 95% RPM:

- ♦Power Available Light **ON** (Blue)
- ♦APU OIL PRESSURE LOW Light out
- ♦Door Sequencing should be complete
- ♦Ignition cuts out
- ♦Timer begins for 60-second bleed air delay

"Normal" APU EGT—200-400°C or 60-70% With **APU AIR** switch ON & all pneumatically powered systems off...QRH, p.3 APU-2

Note: *Only one reset of APU generator permitted for each APU start*
If the APU generator does not reset, contact maintenance and make an E-6 entry ....OM Vol. 1 PRE-FLIGHT 10.6

Autoflight

Editor's Note: *No attempt is made here to describe the autopilot system in detail. That would be beyond the scope of this document. Only key highlights which are easily missed or forgotten will be covered here for review. It is assumed that most pilots interact with this system sufficiently on a day-to-day basis to be familiar with its operation. In addition, the FGS Trainer at the Flight Academy does an excellent job of introducing this system.*

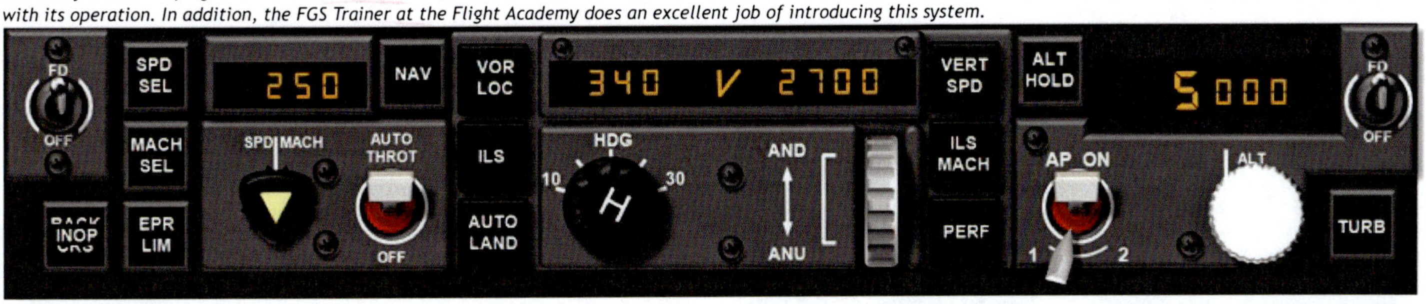

General

Limitations—Review OM Volume 1, LIMITATIONS, pages 10.12-13 for numerous items

Autopilot (AP)

Disconnecting AP

Hold control yoke when disconnecting AP

If trim input is made while AP is engaged:
- AP would compensate
- When AP disengaged, abrupt control input could result

Verbal callout is required when disengaging AP *or*
- if AP is observed to have disengaged itself

OM Vol. 1, SYSTEMS 25.9

Heading Split

±2° or Greater split causes
- Flight director V-Bars to bias out of view
- AP disconnect
- NO AUTOLAND lights illuminate

Above functions return to normal when split is eliminated

If heading is 360° ± 3° at time of split, functions not operational until heading changes from 360° by at least ± 5°

Engine Failure on Takeoff Logic

Loss detected below V_2—V-Bars command and AP controls pitch to accelerate to V_2

Loss detected between V_2 and V_2 + 10—Target pitch is used to hold airspeed constant

Loss detected above V_2 +10—Target pitch is used to slow to V_2 +10 (trade airspeed for altitude)

Engine Failure Logic (AD 92-10-13 R1)

Armed if
- (1) the flight director pitch axis is in takeoff mode,
- (2) the aircraft is above 350 feet radio altitude, and
- (3) both engine pressure ratios (EPR's) are below the go-around EPR limit

If the DFGC detects an EPR drop
- ≥ 0.25 EPR *and* ≥ 7% N1 compared to the other engine *then*
- Engine failure logic is satisfied *and*
- DFGC will change the Thrust Rating Panel (or indicator) thrust limit to Go-Around (GA)
- This will cause the autothrottle system to unclamp and enter normal EPR limit (EPR LIM) mode where the throttles will maintain the higher engine EPR at the selected go-around thrust rating EPR LIM
- Such an EPR and N1 drop may also result from an engine surge (stall).
- Advancing thrust levers on a surging engine will hinder surge recovery and may result in eventual engine failure.

If an engine surge (stall) is detected during takeoff:
- Follow procedures in OM Vol 1 for Engine Stall/Surge after TO

Flight Guidance Differences

FMS OVRD Button replaces **BACK CRS**

VNAV Button replaces **PERF**

See TWA Supplement for other FGS differences.............. TWA

Autothrottle (AT)

Autothrottle Clutch Mechanism

Allows manual throttle movement with AT on, *but*
- Wears clutch mechanism excessively
- Maintenance repair takes extensive time

To avoid, throttles should only be manually positioned with FMA throttle window **CLMP** or **LOW LIM**

CADC Light—See page 40 of this study guide

FD Light—FD CMD Selector is out of the **NORM** position.

Other Systems

Yaw Damper

Activated when:
- Yaw Damper switch is ON
- or
- Yaw damper switch is OFF but Autopilot is Engaged

Deactivated when switch is in **OVRD**

Rudder movements generated by the Yaw Damper are not transmitted to rudder pedals

Autopilot *will* engage with yaw damper deactivated (Switch in OVRD)

Altitude Alert

Deactivated at glide slope capture

Reactivates for go-around (if required) when glide slope signal lost during missed approach

EFIS

Excessive pitch chevrons appear at:
- +25° Pitch
- -7° Pitch

Full deflection on Fast/Slow display equates to approximately 10 Knots

There is *no failure warning* for marker beacons INOP

Note: **If EFIS display goes full bright**—indicates a problem with the dimming circuit which is uncorrectable.

Flight Mode Annunciator

Four Columns of displayed information—*TARP*
- **T***hrottle*
- **A***rmed*
- **R***oll*
- **P***itch*

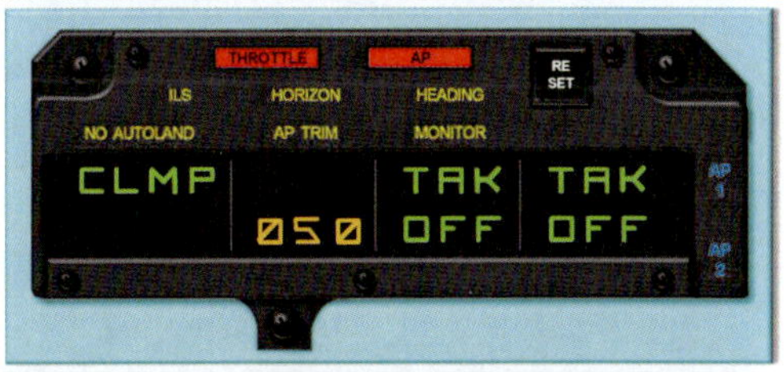

Autothrottle Mode	"Armed" Display	Roll Mode	Pitch Mode

Reset Button
- Allows reset of all the amber lights in the diagram above except the **AP TRIM** indication

Communications

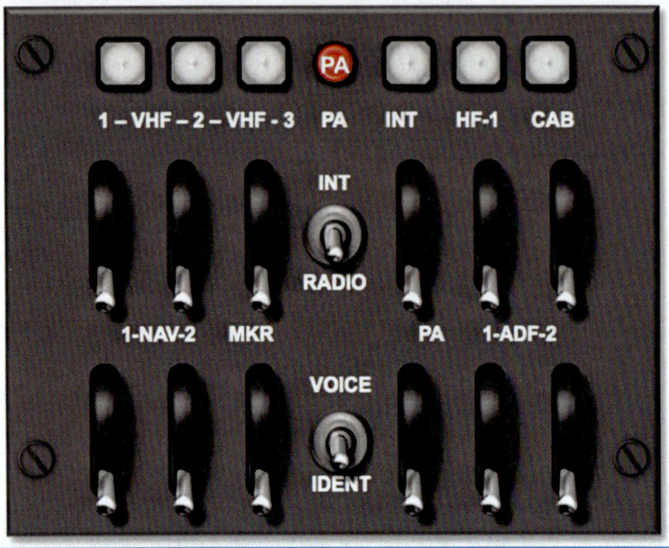

General

Two separate transmitter/receivers for VHF
- #1 Radio—on Emerg DC Bus; normally used for ATC
- #2 Normally used for ATIS, Company, Ramp, etc.

Some aircraft require boom microphones below 18,000 feet MSL

Selective Calling (SELCAL)

Two monitors, each tuned to one VHF radio

SELCAL 1 monitors **VHF 1**

Each looks for coded signals indicating a call to the aircraft

If received, SELCAL cues the Central Aural Warning System (CAWS)
- Light for appropriate SELCAL is lit
- Chime sounds

Chime Reset: Push SELCAL light

Interphone and PA

Microphone selection—Very straightforward design
- Light comes on and button of mic selected remains depressed to indicate which is currently selected
- Only one mic selection may be made at a time

Pushing PA mic selection allows PAs to be made with the O_2-mask on

Cockpit speakers are muted during transmissions from CA or FO positions to prevent feedback

Mode S Transponder

Ground Operation
- Ground control relay inhibits some functions while aircraft is on the ground (*instructor note*)
- "R" Reply indication comes on if ATC 2 is selected and system is interrogated
  - Not required on Mode S transponders
  - Does not function with ATC 1 selected

Some aircraft have two installed, some only one

ATC FAIL Light—Indicates selected transponder (1 or 2) has failed

ACARS

Uses third transceiver

Printer Alert light may be reset by pressing one of two buttons:
- Printer Alert Reset Light on ACARS printer
- Printer light on SELCAL panel

Otherwise, similar to ACARS on other aircraft except:
- No Takeoff Power Used reports need to be made
- Engine Monitor Logs are automated
- Closeout sent automatically with OUT event
- See ACARS Quick-Reference Cards

PRINTER MESSAGE Light indicates a paper message has been delivered

PRINTER MESSAGE

Cabin Interphone Button

Connects oxygen mask microphone to Flight Attendant's cabin interphone system

Button is installed on communications panel

Electrical Systems

General

Three essentially identical 40 KVA Continuous output generators: One per engine; One on APU

Generators are never paralleled

DC power through 4 Transformer-rectifiers
- Any 1 of these can supply entire airplane load
- 2 Powered by Left AC Bus
- 1 Powered by Right AC bus
- 1 Powered by Ground Service Bus, which is normally powered by the Right Generator Bus

Automatic AC Crosstie connects the two AC Buses if only one is powered by an engine-driven generator unless a bus fault is detected

Manual Crosstie for DC system

Two 14V Batteries in series to provide 28V DC

Emergency Inverter
- Powers AC Emergency Bus by converting battery DC to AC if normal AC fails
- Same inverter powers refueling system when normal electrical power not available

AC Power

Generators

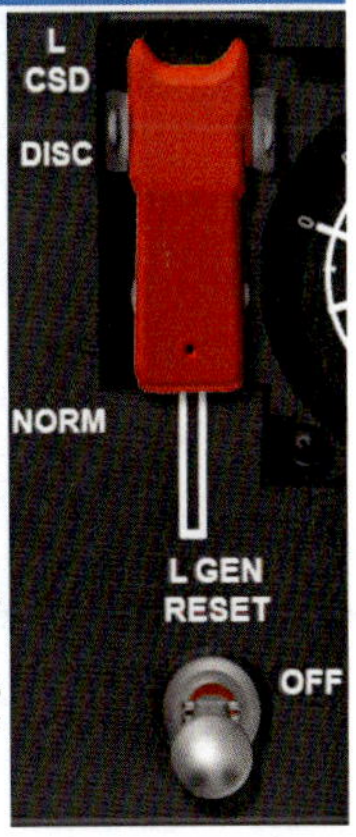

Constant speed drive maintains constant frequency output
- Can monitor CSD Oil Outlet temperature and temperature rise across the drive
- Monitor outlet temperature in flight—have to push and hold a button to check rise

Fault Protection
- Generator is removed from its bus and de-energized for "certain faults"
- Faults not listed in Flight Manual, and you can't fix them, so no list to learn!
- Fire handle trips respective generator control relay
- Can reset respective generator after fault has cleared (after fire handle reset, if applicable)

Priority for Power Selection in Automatic Switch Positions

Only the highest priority power will be taken, in the following order:

- ❶ Engine Generator
- ❷ APU
- ❸ External Power
- ❹ AC Bus Crosstie

Example: APU is powering all buses when left engine is started.
- When left engine comes up to speed and engine generator power is stable, APU generator will be replaced by the left engine generator in powering the left AC and (through TRs) left DC buses
- Shutdown of a generator automatically transfers its buses to the other operating generator through the **AC Crosstie Relay** (unless bus fault detected, then the bus remains unpowered.)

AC Crosstie relay

Automatically connects one Generator to opposite bus if power source is disconnected

Two Exceptions:
- Bus Fault detected (i.e., short on the bus)
- Differential fault protection is activated

Protects operating generator from being damaged

Loss of L or R AC power activating AC Crosstie disconnects Galley Power *automatically*

APU will crosstie *in flight only*

External power *never* crossties

Power Distribution

Normal—Generator connected to its respective bus

AC Crosstie allows generator to power both sides

Only one generator can be selected for a given bus

APU generator can power either or both buses

Emergency AC Bus
- With the Emergency Power Transfer Relay unpowered, spring-loaded to the ***right*** side
- When connecting APU power—Left side first; check DC EMER BUS OFF light out
- Ensures power switches automatically to the ***left*** side when only the left side is powered
- Normal power source Left side when powered

Ground service bus—powered by APU or external power
- To be powered by External power, APU must be disconnected from the bus first; ***if not:***
- Indication—AC CROSSTIE LOCKOUT Light
- If lockout fails—GS bus powered by opposing sources, which can overload each other or the bus

Power Available Lights

Provided for APU and External Power—both indicate same for respective source

APU Power Available light *and*
External Power Available Light
- Power is available
- Power meets frequency/voltage requirements
- Phase sequence is correct
- However, ***NEVER*** a bad idea to check power before connecting to airplane

Ground Service Bus

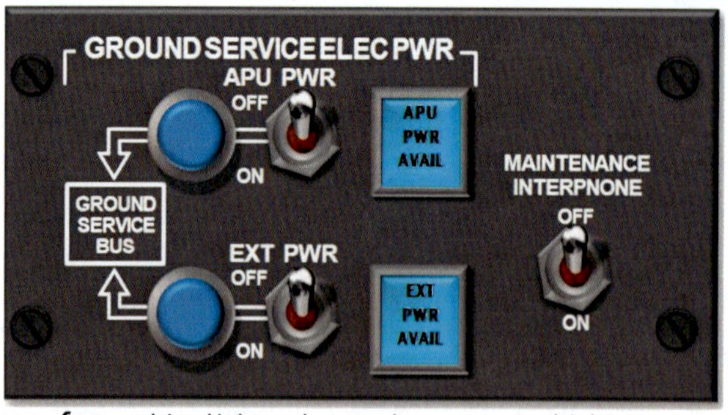

Power for cabin lights, lavatories, etc., which are used to service the airplane at the gate

No power to avionics

Normal power source is from right AC distribution system with all buses powered

Can be powered from APU or **external power** without other airplane buses being powered

Key Operational items on **Ground Service Bus:**
- ♦Right Aft Boost Pump
- ♦Battery Charger
- ♦Center Instrument Flood Lights

DC Power

DC Bus Normal Power Sources

Left DC side is separated from right DC, similar to AC system

DC buses (L & R) are cross tied ***manually*** rather than automatically

Left side power
- Left TR 1 & TR 2 are powered by L AC bus
- TR 1 & TR 2 load share to power L DC bus

Right side Power
- TR 1 is powered from right AC Bus
- TR 2 is powered from the Ground Service Bus, which is normally powered by the right generator bus when all buses are powered
- Allows battery to be charged when ground service bus is the only bus being powered through external power on the ground

Emergency DC Bus

Normal power source:......................................Left DC bus
Left DC Bus unpowered:........................source is Right DC bus
Both DC buses unpowered:.............source is Battery Direct bus
(Power source then is the Battery)

Emergency Power Selector

Connects Battery to Emergency AC & DC Buses

Disconnects normal power from these buses
- Fully-charged Battery should be good for 30 minutes with normal in-flight loads
- Normal battery load carrying these buses is 10-30 amps (up to 50 is allowed)

White EMER POWER IN USE light confirms battery is powering these buses

DC Transfer Bus Off Light (Amber)

Indicates no power to this bus

Should be off when battery is first turned on—indicates battery is powering DC Transfer Bus

After Generator and TRs are on line, the bus will be powered by these sources

Key items on DC Transfer Bus (SAFEE)

Standby Horizon

APU Door Control

Fire—Detection & Protection
- Agent Discharge Bottle Arming
- NOT Agent Low Lights
- Can't confirm bottle discharge with only Battery & EMERG buses powered

Engine Start Pump

Engine Ignition Override

Ni-Cad Battery & Battery Charger

Battery Charger Light—Indicates battery charger failure

Pre-Flight Check
- During charging, amps read approximately 40
- Reading decreases as charging completes
- Reads 0 (*zero*) when fully charged
- Normal limits can be verified by:
  - Meter Selector to BATT AMP
    - Needle should be centered or to left, indicating charging
    - Needle should not be displaced right, indicating discharge
  - Meter selector to BATT VOLT
    - Verify voltage in normal limits of 29±4 Volts DC

Miscellaneous

Volt/Amp Indicator Selector

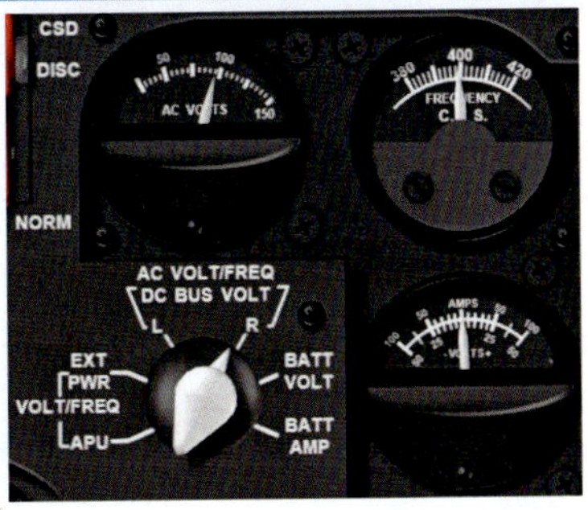

L or R positions
- AC voltage & frequency for selected sources on respective meters

Other positions
- Charge or discharge of current battery, battery voltage, or DC Bus voltage
- BATT/VOLT Position: Bottom scale reads voltage
- BATT/AMP Position: Top scale reads amperage
  - Needle displaced right indicates battery discharge
  - Needle displaced left indicates battery charging

Quick Reference for Which Buses are Powered

Captain's instruments	AC Emergency Bus
Captain's Pitot Heat	DC Emergency Bus
Standby Horizon	DC Transfer Bus

DC Bus Backup Power Sources

The outline on the following page is cumulative—it lists buses powered at each point in the sequence. All the buses above each level remain powered throughout the remainder of the sequence.

Battery OFF—
- Battery Direct Bus

Battery Switch ON
- Battery Bus
- DC Transfer Bus

Emergency Power Selected
- Emergency DC Bus
- Emergency AC Bus, through Emergency Inverter

Other Warning Lights

APU Generator Off Light MC
- APU is Operating and
- APU Generator is not on line

L or R GEN OFF Light MC
- Generator relay is open
- L/R Gen disconnected from bus

BATTERY OFF light
- Indicates **BATT** switch is in **OFF** position

AC Crosstie Lockout Light
- AC Crosstie is locked open
- Automatic crosstie inop

DC Bus Off Light MC
- L or R DC bus is unpowered
- Or DC bus sensing circuit open

AC Bus Off Light (L or R) MC
- Respective Bus is unpowered
- Or AC bus sensing circuit open

CSD Oil Pressure Low (L or R) MC
- CSD Oil Pressure is below operating limits

Power Loss Chart

Condition: No Generators On Line; Battery Switch ON; Emergency Power Selector OFF

Buses Powered: DC Transfer Bus; Battery Bus; Battery Direct Bus (Powered even with **BATTERY** Switch **OFF**)

Capability:
- ♦Maintain Attitude
- ♦Fight Fires
- ♦Start Engines
- ♦Evacuate Passengers (All of these for 30 minutes)

Instruments & Radio	DC Transfer Bus
Standby Attitude Indicator	X
DC Start Pump	X
Override Ignition	X
Master Warning/Caution lights; Annunciator lights) MC MW	X
Oil Pressure Low Caution Lights	X
Fire Detection and Protection (no AGENT LOW lights)	X
APU Door Control	X
Generator Control	X
Pneumatic Airspeed & Altimeter	
Mag Compass	
N1 & N2 Indicators (Self-Generating)	

Lighting
- Cockpit flood lights
- Passenger cabin dome lights
- Passenger emergency exit lights
- Seat-mounted emergency escape path lighting

Condition: No Generators On Line; Battery Switch ON; Emergency Power Selector ON*

Buses Powered: Three Above Plus…; Emergency DC Bus; Emergency AC Bus**

*Turns off Battery Charger
**Through Emergency Inverter

Capability: Navigate / Communicate / Penetrate / Shoot an ILS; Tell Passengers you are going to Evacuate Them

Still limited to the Battery life duration (30 Minutes)

Instruments & Radio

Captain's
- ♦Attitude Indicator
- ♦IVSI
- ♦Electric Altimeter
- ♦Electric Airspeed Indicator
- ♦HDI & Glide Slope

F/O's RMI and #1 Needle

Other Powered Items
- ♦#1 Navigation Radio (& ILS for Cat I approach—250' and ½ mile visibility)
- ♦#1 Needle, not Card, on RMDI
- ♦#1 Comm
- ♦VG Switching Capability
- ♦GFMS
- ♦Intercom
- ♦EGT Gages
- ♦No Flight Director / Radio Altimeter / DME
- ♦Public Address

Lighting
- Cabin Standby Lights
- Captain & F/O Variable Panel Background Lighting
- If activated, the following are also lit with power from rechargeable battery packs for approximately 15 minutes:
  - ♦Passenger aft stairway and tail section lighting
  - ♦Exterior over wing emergency evacuation lights

Other
- Air Conditioning Control in Manual; Pack Control (ON/OFF)
- Cargo Smoke Detectors Loop B
  - Loop A is DC Transfer Bus
  - Loop B is Emergency DC Bus—See OM Volume 2, page FIRE PROTECTION-2)
- Hydraulic Pumps revert to high output (no indication)
- Manual pressurization
- Flight spoilers will deploy to 60° manual limit
- *No Ground Spoilers (Electrical operation)
- *No flight directors or radio altimeters
- *No Trim
- *No flap indications, gear or slat lights
- *Engine Anti-Ice remains as previously selected
- *Wing Anti-Ice fails to OFF
- *No Anti-Skid
- *Cockpit door—Operative, but must be manually secured with deadbolt

*(Notes from ground school—not in OM)

QRH Tabs page 9.5-6

Engines

General

Engine Designation & Thrust

Pratt & Whitney JT8D

Engine	Normal T/O Thrust	Max T/O Thrust
-219	21,000 lbs.	21,700 lbs.

Oil System

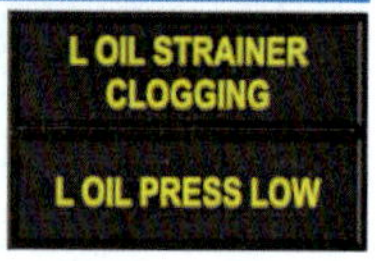

Engine-driven main oil pump delivers oil to engine

OIL STRAINER CLOGGING Light **MC**

Excessive differential pressure at oil filter

L or R OIL PRESS LOW lights **MC**

Fuel System

Fuel Path to Engine:

♦Engine-driven, first stage Centrifugal Pump

♦Air-Fuel Heat Exchanger

13th Stage bleed air used to heat fuel

♦Fuel Heat Switch controls

Automatic 1-minute timer

L or R FUEL HEAT ON Light indicates system ON (*Blue*)

♦Fuel Temperature Sensor—Downstream of heater

♦Fuel Filter

L or R FUEL FILTER PRESS DROP Light **MC**

Could be ice or contamination *but*

Ice is unlikely, especially when fuel temp is above 0°c

Bypass valve allows fuel to pass a clogged filter

♦Fuel Flow Transmitter is between fuel control and fuel-oil heat exchanger

♦Fuel-Oil Heat Exchanger

Cools oil, preheats fuel

Only reference—schematics, OM Vol. 2, ENGINES-20.3, 20.5

Thrust Reversers

Clamshell Doors can easily scrape on runway if deployed before nose wheel touchdown on landing

Two Status Lights:

ENG REVERSE THRUST Light—Associated reversers fully extended

ENG REVERSE UNLOCK—Associated Reverser unlatched

One Warning Light

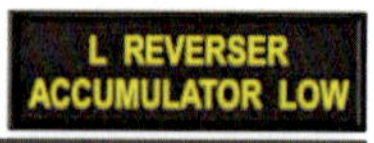

L (or R) Reverser Accumulator Low

Amber; no Master Caution Light

Starting & Ignition

Starting Power

High pressure air

Sources possible:

♦APU

♦External Air Cart

♦Opposite engine, using crossbleed*

*If used, need ≈36 psi, or N_2 80% .... OM Vol I, START-15.8

Starter

♦Electrically controlled

♦Pneumatically activated

L or R START VALVE OPEN Light for each engine on overhead panel indicates start valve butterfly is open

Ignition

2 Igniters per engine

Power Output:

High-energy System—20-Joule output

Low-Energy System—4-Joule output

Ignition Switch

OVRD—

High energy ignition to both igniters

No regard to position of start switches or fuel levers

CONTIN—

Low energy output to one igniter per engine

Fuel lever must be ON for respective engine

TWA A/C Ignition Switch

Has A and B systems

A or B Selected

If fuel control lever ON:

20-Joule AC ignition from respective (A or B) igniter

BOTH Selected

If fuel control lever ON:

20-Joule AC ignition from both (A and B) igniter

OVRD Selected

20-Joule AC ignition from both (A and B) igniter

Fuel control lever is bypassed .... **TWA**

Start Switch

Three Position Switch

GND only—Start Valve opens if air pressure available

GND or FLT—

High energy ignition to both igniters

Fuel lever must be ON

No regard to position of ignition selector

Spring-loaded to OFF

TWA Aircraft Start Switch

OFF Removes power from engine start valve

ON Provides power to open start valve

No ignition function............. **TWA**

Automatic Reserve Thrust (ART) System

Basic design:

Automatic detection of engine failure during takeoff

Looks for one N_1 30% below the other (*See added notes p. 26*)

Thrust increased on operating engine at the fuel control

ART activates on max thrust takeoff with windshear encountered (OM Vol. II, p. WARNING & ALERT-50.4)

ART INOP Light on overhead panelART fault detected or system OFF with above conditions met

ART Self Test:

Initiated automatically when **(NAGS)**

N_1 Signal—Both engines running

Auto position for ART switch

Ground—Aircraft on the ground

Slats Not Retracted (Mid-sealed or fully extended)

Self test complete—READY light (green) on—no faults detected, system operational

ART System Arming

Self test complete (READY light—see above)

Both engines above **64%**

System is then ARMED and looking for 30% N_1 difference

ART Light indicates when............................... ENG-10.9, 20.8

♦Engine Failure has occurred (N_1 input) *or*

♦DFGC switched (1-2) or failed *and*

Reserve system has activated (results-see following)

ART System Actions when Triggered
ART solenoid shifts fuel control to a thrust schedule that increases EPR
Equates to approximately +.05 EPR
Operating engine instruments show increase in N_1, N_2, EPR, & Fuel Flow (No movement on throttles)
TRI updates to computed reserve thrust setting
Note: Thrust may increase above TRI-indicated setting. Do not retard throttle to match.

ART System Disarming After Takeoff
Before Slat Retraction
Both engines retarded below 58% N_1
READY Light remains ON
Re-arming occurs when power on both engines is advanced above 64%
After Slat Retraction—
Permanent disarm
Subsequent re-arm occurs after landing

ART Switch
AUTO—Automatic reserve thrust system enabled
OFF—ART system disabled

Automatic Thrust Restoration (ATR) System

Basic Design
Senses performance loss after takeoff
Assumes engine loss has occurred
Works through DFGC to unclamp throttles and increase power on working engine to Go-Around Power

Arming—TAP
T*akeoff*—Flight director in take-off (TAK OFF) mode
A*ltitude*—Radio Altitude over 350 feet
P*ower*—Both engines operating below Go-Around power
Note: *If engine failure occurs below 350 feet, ATR does not engage when altitude reaches 350 feet. (See added notes* *page 28*)

Activation
Vertical speed decrease to less than 0 fpm for five seconds *OR*
EPR drop ≥0.25 *and* N_1 loss ≥7% on the same engine

Actions Occurring when Activated
Throttles are unclamped (if ATs are engaged)
Both throttles advance equally until one of the engines reaches Go-Around EPR limit
If ART system is *armed* and ATR *activates*
ATR Target thrust limit is reduced by the amount the ART system is scheduled to provide
Prevents over-boosting with both systems attempting to augment thrust on same engine

Disarming
Occurs when any pitch mode selected on DFGC
Caution: *DFGC can incorrectly sense engine surge as engine failure.*
♦Advancing throttle in this case worsens situation.
♦Auto-Throttle limit requires disconnecting AT if surge or compressor stall occurs on T/O for this reason.

Other Systems

Approach Idle
≈10% Higher than normal idle
Causes a slightly higher than normal idle to assist in the event of a go-around and need for engine spool-up (with or without engine failure)
Begins—5 Seconds after nose wheel indicates down
Ends—5 seconds after nose strut compression

Thrust Rating Indicator (TRI)
Interfaces: ♦RAT Probe ♦**Both DFGCs**
Reserve Thrust Selection ♦TRI window—00° Selected
♦ART Switch OFF ♦T. O. Button depressed
Max Takeoff Thrust Selection ♦TRI window—00° Selected
♦ART Switch ON ♦T. O. Button depressed
Uses RAT temperature to calculate max thrust
Engine failure detection by ART triggers display of Reserve Power on EPR gage of operating engine (*See added notes p. 26*)

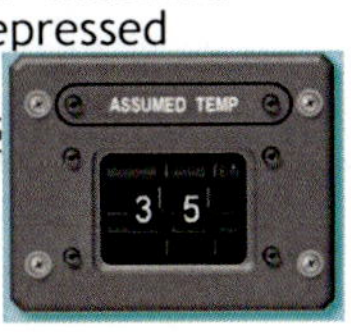

Standard Power Takeoff Selection
♦TRI Window—Assumed temperature selected
♦**ART** Switch **OFF**
♦**T. O. FLEX** Button depressed
Assumed temperature input used to calculate thrust

Takeoff Sequence
T. O. or T. O. FLEX buttons depressed
EPR bugs display EPR Limit as selected
After 50 knots A/S, computer stops recomputing EPR
At 60 knots autothrottle clamps if engaged

TRI Flight Modes & Indications
"Bottom Row" Buttons
♦**CL**—Climb
♦**MCT**—Max Continuous Thrust
♦**CR**—Cruise

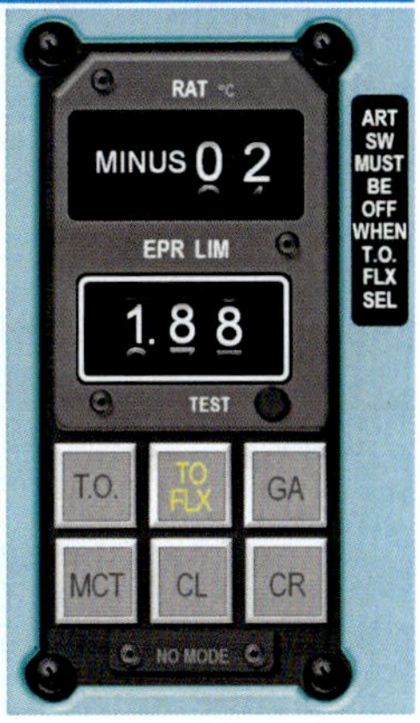

NO MODE Light indicates
EPR Limit mode is not selected *or*
Improper configuration for flight conditions (uncertified engine bleed configuration—see below)
♦Airfoil AI **ON**; engine AI *not* **ON**; pneumatic crossfeed lever(s) in *any position*
♦TO or TO FLX modes, engine AI ON, RAT>10 C
♦GA mode, engine AI ON, RAT>14 C
In other words, engine AI on >10°C ground/14°C flight
♦MCT selected, airfoil AI ON, both X-Feeds OPEN
♦CL selected, airfoil AI ON, and one pneumatic crossfeed lever CLOSED
♦MCT, CL, or CR selected with packs off
♦T. O. FLX selected, ART switch AUTO
In addition to light, EPR Lim flag appears

Engine Synchronizer System
Trims left engine to right engine
EPR is automatically synchronized with:
♦Engine sync selector OFF
♦Autothrottle engaged
♦Any autothrottle mode except CLMP
With Engine Sync Switch in N_1 or N_2
Matches N_1 or N_2 if they are within ±1%

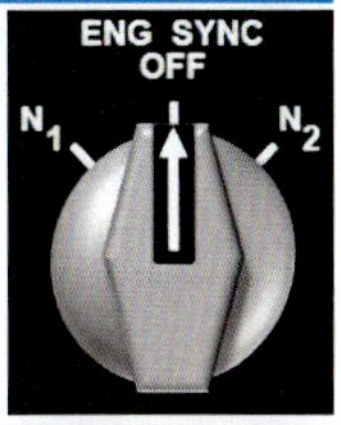

ENG SYNC ON Light
Landing gear handle down
Engine Sync Switch in N_1 or N_2

System should *not* be on for
♦Takeoff
♦Thrust reverser operation
♦Landing
♦Airplane below 1500' AGL
Prevents problems, especially in the event of engine failure during a critical phase of flight

Engine Instrument Power Sources
AC Lies, DC Dies
AC-powered instruments freeze where they are
DC-powered instruments drop to zero
Categories:
All RPMs are Self-Generating (N_1, N_2)
All temperatures are—*DC* (EGT, Fuel, Oil)
All direct reading items—*DC* (temperatures oil quantity)
All pressures are *AC* (EPR, Oil)
All computed indications are *AC* (EPR, Fuel Flow)

Summary:

EPR	AC	Fuel Temperature	DC
N1	Self-Generating	Oil Temperature	DC
N2	Self-Generating	Oil Quantity	DC
Fuel Flow	AC	EGT	EMER DC
		Oil Pressure	AC

Engine Display Panel (EDP)
Engine data displayed on digital center screen
TWA 94xx and 96xx airplanes are affected
See TWA Supplement for details .... TWA

Throttle Response Failure
See "EPR Erratic or Fixed / Displays Dashes" procedure
QRH ENG-12

Fire Protection

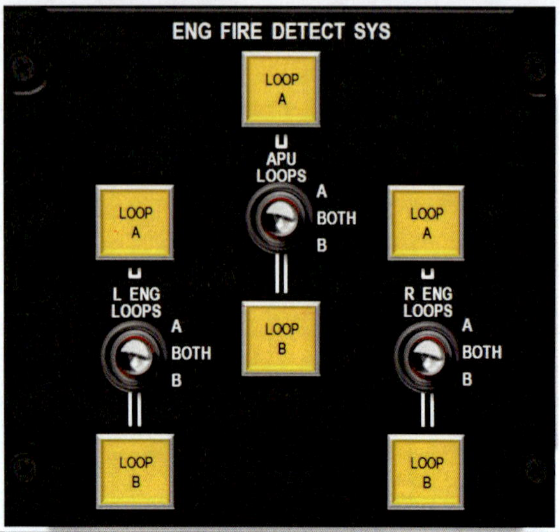

Fire Detection Systems

Fire Loops—6 Total

Two Loops on each engine and the APU
Loops "A" and "B" for each of the 3 systems

Fire Loop Logic

Loop Switch in BOTH—
No fire warning unless both loops detect fire
Fire Warning Light in Both A & B Loops
Components activated when both loops detect fire

ENG FIRE light in engine fire handle, forward panel
Separate lights for **LOOP A** and **LOOP B**
Fire Detector Loop caution light
MASTER CAUTION Light **MC**
FIRE DETECTION LOOP caution light

Fire Bell
Aural Warning: "*FIRE LEFT ENGINE*"
Pressing FIRE BELL OFF button
Silences Bell
Silences repeating aural warning
Fire detected in only one loop:
Loop light for respective loop
MASTER CAUTION Light **MC**
FIRE DETECTION LOOP caution light

No aural warning
Loop Switch in A or B
Fire detected in that loop generates same result as above for both loops detect a fire with switch in BOTH

APU Fire Detection

All of the above details concerning engine fire loops apply to the APU *except*:
- **Aural Warning is:** .... Alternating horn and "*APU FIRE*"
  Repeated three times
  Cannot be silenced
- **External Horn**
  Lower aft body, under left engine
  Will not silence until the fire is out
  Does not sound during test
- **APU FIRE Light** **MW**
- **APU Fire detected** arms external controls for fire bottles
- APU Fault Light (See *page 27 of this Guide* for detail on this light)

Fire Protection Systems

Pulling/Turning Engine Fire Handles

Pulling Fire Handle—Five Actions result
2 Electrical effects
2 Fluid-related effects
1 Air-related effect
Turning Fire Handle—One Action Results
1 Fluid-related effect
Power Sources
Electrically activated actions from DC Transfer Bus
Mechanical actions are from a cable connection

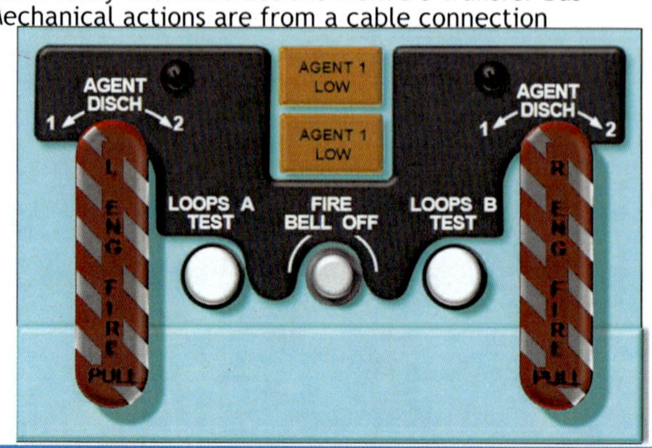

Summary:

Pulling Fire Handle:

Power	Action	Category
Electrical	Aural Warnings Stops	Electrical
Electrical	Generator Control Relay Trips	Electrical
Cable	Fuel Shut off at Tank	Fluid
Cable	Hydraulic Pump Supply Shut Off	Fluid
Cable	Pneumatic Crossfeed Valve Closed	Air

Turning Fire Handles:

Cable	Fire Bottle Discharges	Fluid

APU Fire Protection

APU Fire switch to **OFF/AGENT ARM**
Shuts off Fuel to APU
Arms APU FIRE AGENT switches for discharge of bottles into APU
Generates overspeed signal to shutdown APU, bypassing 60-second timer
Trips APU Generator Relay (*Instructor note, not in OM*)
External APU SHUTOFF switch accomplishes the same actions as above

Agent Low Lights

Confirm Discharge
NOT Powered by any of the 5 battery-operated buses
If battery is only power source—lights inoperative

Lavatory Smoke Detectors

Purpose

Designed to detect and extinguish trash bin fire
May not detect smoking in lavatories

When Activated by Smoke Detection:

Cabin Lights Activated:
Associated Lavatory Call Light
Master Call Light
Lavatory Chime sounds at 1-2 second intervals
No integral alarm comes from the detector itself
Dissipation of smoke
Causes signals to stop
System resets automatically

Cargo Smoke Detection & Fire Suppression(SDFS)

Main Functions (When installed)

Detection of smoke in FWD, MID, and AFT compartments
Alerting of crew to presence of fire/smoke condition
Suppression of fire in one cargo compartment

Components

Cargo Fire Panel (overhead panel; replaces APU hr. meter)
CARGO FIRE light on overhead annunciator panel
Circuit breakers for loops A and B

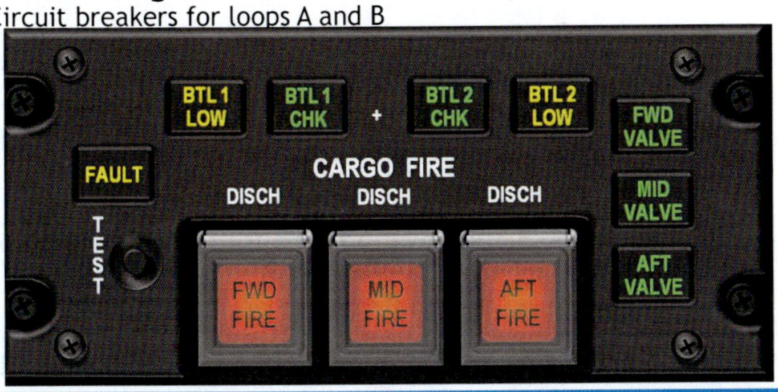

Detection

Photo-electric cells and particle detectors in each cargo compartment
Each cargo department divided into several zones
Each zone has **two detectors** (one for Loop A, one for B)
Detection occurs when both detectors in a zone detect smoke
If one detector in a pair has a sensed fault, the remaining detector alone triggers detection if smoke is found
Smoke detection occurs "within 60 seconds"

Crew Alert

MASTER WARNING Light on glare shield
CARGO FIRE Light on overhead annunciator panel
Associated FWD, MID, or **AFT FIRE** Light/switch on Cargo Fire panel

TWA **Aural tone** sounds
Tone is silenced by momentarily pressing **ALARM OFF** button on cargo fire alarm panel

Suppression

Two halon fire bottles installed on right side of mid cargo compartment
Halon directed to proper compartment by diverter valves
First bottle—when compartment light/switch is pressed:
- Discharged immediately, at high rate
- **BTL 1 LOW** Light indicates bottle has been discharged

Second bottle
- Discharged automatically after 15 minutes, at a slower rate
- **BTL 2 LOW** Light indicates bottle has been discharged

Lights and Switches

FWD, MID and AFT FIRE Lights
ON—when:
- ◆Both detectors in the same zone detect smoke ***OR***
- ◆One detector has faulted ***and*** the other detector in the same zone detects smoke

Pressing illuminated cargo compartment fire light discharges halon into compartment (see above)

FAULT Light—On when:
- Test Switch pressed
- SDFS Failure detected
- If test is done with a valid fault present, light goes out after test and comes back on after 2-second delay

BTL 1 (or 2) LOW Light—On when:
- Test Switch pressed
- Bottle pressure low—should occur 10-20 seconds after illuminated fire switch is pressed (bottle 1) and shortly after the 15-minute delay (for bottle 2; see above)

BTL 1 (or 2) CHK Light—On when:
- Test Switch pressed
- Verifies continuity of respective squib

FWD, MID and AFT VALVE Lights On when:
- Test Switch pressed
- Verifies proper function of corresponding halon diverter squib

Cautions: ***Whenever squib lights are illuminated****, even during test, pressing a bottle discharge switchlight,(i.e.* **FWD FIRE**, *etc) will discharge halon.*

Resetting **MASTER WARNING** *Lights during test will disable Cargo SDFS from illuminating* **MASTER WARNING** *Lights again. Pull and reset c-b on overhead c-b panel (A19 and A20), if required.*

TWA Cargo Smoke Detection & Fire Control

Detection—Similar to AA aircraft
A and B Loops selectable
- **A** or **B** Position—enables only that loop
- **BOTH** position is the normal mode
  - Arms both loops for smoke or high temperature detection

Protection
- Exists in any cargo compartment
- Cargo fire warning automatically arms number 1 fire bottle
  - **BTL 1** Switchlight illuminates, indicating armed
  - To discharge, crew must raise guard and push **BTL 1** switchlight
- Discharging bottle 1 arms the number 2 fire bottle
  - **BTL 2** Switchlight illuminates, indicating armed
  - Bottle 2 discharges automatically after 15 minutes if still airborne

System is powered by DC Transfer Bus .... **TWA**

Additional Details—See:

FIRE Warning Test ....OM Vol. I, PRE-FLIGHT 10.20
TWA **Cargo Smoke Detection & Fire Suppression System (SDFS)** ....OM Vol. I, PRE-FLIGHT 10.20

Flight Controls

General

Primary Flight Controls
- Ailerons
- Elevators
- Rudder

Secondary Flight Controls
- Leading Edge Slats
- Trailing Edge Flaps (Inboard & Outboard)
- Flight and Ground Spoilers
- Horizontal Stabilizer

Most Primary Flight Controls are moved via cable-driven control tabs

Most Secondary Flight Controls are hydraulic

Primary Flight Controls

Lateral Control

Ailerons

Control wheels
- Connected to aileron control tabs via cables
- Control wheels are linked together by torque tubes
- Control Tab "flies" aileron to the necessary position

If part of the system jams:
- Either control wheel will drive unjammed portion and free control tab
- Approx. 50 pounds of force at wheel rim will push one side with other side jammed

Lateral Control Augmentation

Trim Tab
- Located on aileron trailing edge outboard of control tab
- Trim knob is cable-connected to trim tab

Flight Spoilers
- Deploy in proportion to aileron deflection
- Initial deployment is at 5° control wheel movement

Longitudinal Controls

Elevator Control—Via Control Tabs (Inboard)

Each control wheel is cable-connected to a control tab on its respective elevator

Only connection between control wheels (and therefore the two elevator sides) is a torque tube between the control stick on each side of the cockpit

If one control wheel cable breaks, either control wheel remains connected to the remaining elevator control tab

Elevator "Geared Tabs" (Middle)

Outboard of control tabs

Move to assist control tab movement

Geared to *elevator* movement

Elevator Anti-Float Tabs(Outboard)

Outboard of geared tabs

Geared to *Stabilizer* movement
- Trim motors reposition Stabilizers (See Below)
- Improves longitudinal trim in forward center of gravity (beyond 10° airplane nose up) landing configuration

Stabilizer Trim

Three methods of control input
- **LONG TRIM** Handles (Longitudinal trim)
  - Primary Trim motor
  - Fast trim
- Control wheel trim switches
  - Same primary trim motor as above
  - Fast trim
- **ALT LONG** Trim Control
  - Alternate trim motor
  - Slow trim

Priority is in the order above
- LONG TRIM Handles override input from yoke switches or alternate switch inputs
- Control wheel switches override ALT LONG trim switch inputs

Either motor drives a jackscrew, which repositions the stabilizer angle
- Primary trim motor—$\frac{1}{3}$° per second
- Alternate trim motor—$\frac{1}{10}$° per second

Each switch is a combination of two switches
- Motor Switch
- Stab trim brake release
- Neither switch alone can effect trim movement
- Any motor switch combined with any brake release switch will activate trim
- Opposite inputs from opposite yokes at the same time cancel each other

Autopilot uses Alternate Trim Motor

Double generator failure—no trim is available

Trim Aural Warnings

Warning horn sounds for one second
- Sound starts after ½ to 1 degree (approx.) of movement of trim by any of the trim motors
- Tone then repeats every ½ degree of motion

Stabilizer moved by the autopilot > 2° in 30 seconds results in *Aural Warning*: ....***"STABILIZER MOTION"*** (OM Vol. 2, p. WARNING & ALERT 10.15 and FLT-C 20.8)

Trim Motor Overheat Protection

Large motor on primary trim inputs

Too many rapid, short inputs or a few long ones will overheat motor

Motor cuts itself out when it overheats

Reset occurs when the motor cools

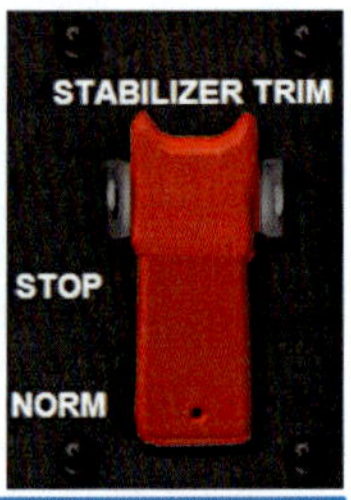

Stabilizer Trim Stop Switch

Stops primary trim motor

Applies stab trim brake

Engine Strakes

Enhance post-stall recovery longitudinal control

Unheated

Elevator Augmentation

Designed to assist in stall recovery

Activation
- Elevator control tabs displaced 10° nose down with respect to the elevator
- Idea—Input must have been applied rapidly, or elevator would have flown to a new position

3000 psi. Hydraulic accumulator
- Charged by left hydraulic system
- Allows significant assist, even when hydraulic pumps are switched to low output

ELEVATOR POWER ON Light

- Indicates hydraulic pressure has been used to assist elevator movement
- Checked on flight control checks

Mach Trim Compensator

Trim spring on F/O's yoke
- Indicator rod on F/O yoke deploys as system engages
- Gives visual verification of system activation

Engages at high mach (above .80 mach)

Pulls yoke aft as mach increases to compensate for airplane's mach tuck at these speeds

Can be overridden with switch on overhead panel

Mach Trim Compensator Indicator Placard
- Indicates direction of movement only
- Can't be used to verify position of compensator

Mach Trim Compensator Inop Light

- Compensator monitor has deactivated the system ***or***
- Mach Trim Comp switch is in **OVRD** position

Yaw Control

Rudder System—Powered Operation

Normal mode of operation is powered

Control tab is hydraulically locked flush with rudder surface

Trim Activation
- Trim knob on control pedestal
- Cable-Connected to actuator to bias the entire rudder surface to either side

Rudder System—Unpowered Operation

Activation—one of two ways:
Rudder Hydraulic Control handle—**MAN**
Hydraulic pressure 950 psi. or less

Either activation
Unlocks control tab to fly the rudder (Normally locked in powered mode)
Rudder pedals move rudder control tab
Trim is still available via trim knob
Restriction with manual control—Minimum speed 135 knots or approach speed until landing assured

RUDDER CONTROL MANUAL Light—Indicates no hydraulic pressure at rudder actuator

Note: *During engine out go around, may require up to 8° of bank in direction of good engine to maintain constant heading at go around speed of V_{REF} + 5 knots minimum or 135 knots, whichever is higher. ....QRH, FLT-C 16*

Trim is available by cable
Connected to rudder control tab
Makes a new neutral position

Nose Strakes

Assist directional control in high AOA flight
Heated by airfoil anti-ice system

Rudder Throw Limiter

Protects empennage from damage with excessive rudder use
Rudder Limiter pitot probe on vertical stabilizer senses speed for input to system

RUDDER TRAVEL UNRESTRICTED Light

Indicates full rudder authority is available
Approach—must be on by 144 knots (MD-82) or 165 knots (MD-83)
If not, restrictions include:
- ♦Minimum speed 135 knots or approach speed until landing assured
- ♦12 Knots crosswind maximum allowable
- ♦During engine out go around—up to ≈ 8° of bank in the direction of the good engine may be required to maintain constant heading at V2 speed or 135 knots, whichever is higher

Takeoff—Must be *out* by 180 knots (MD-82) or 200 Knots (MD-83) ....*See abnormals, QRH, FLC C-16*

Protection is proportional
22° Rudder displacement maximum
Available at approx. 180 knots
Rudder Travel Unrestricted-indicates this situation
≈2½° Available at ≈300K

Light activation indicates full travel (22°) is available

Rudder Limiter Lock: Occurs with full rudder travel from any combination of trim and pedal inputs ***if***:
Rudder Travel Unrestricted Light not yet on
Rudder travel is then limited to number of degrees travel when limiter lock was engaged
Normal operation (rudder limiter adjusting for speed conditions) can be re-established by momentarily:
Centering rudder pedals ***and*** centering rudder trim

Yaw Damper

Damps lateral oscillation
Activated when:
Yaw Damper switch is **ON** *—or—*
Yaw damper switch is **OFF** but AP is Engaged

Deactivated when switch is in **OVRD**
Rudder movements generated by the Yaw Damper are *not transmitted* to rudder pedals

YAW DAMP OFF Light

Yaw Damper is not operating
Yaw Damper switch is **OFF** or **OVRD**

Secondary Flight Controls

Spoilers

General

Inboard & outboard flight spoilers
Operational at all times—In flight & on the ground
Inboard flight spoilers—Left hydraulic system
Outboard flight spoilers—Right hydraulic system

Ground spoilers
Power from both L & R hydraulic systems
Deactivated in flight
Inboard-most spoiler panels
Turbulence would disrupt engine intake airflow

Flight Spoilers

Supplement ailerons for lateral control
5° or more control wheel input—proportional spoiler deployment on downward wing

SPOILER DEPLOYED Light

Ground— Indicates spoiler lever full forward and any spoiler not stowed
Inhibited with TO power set
In Flight—Either ***ground*** spoiler extended

Auto Spoiler Do Not Use Light

Malfunction detected in ground spoiler ***or***
System armed, only one main gear senses ground contact
Does not preclude use of manual spoilers

Spoiler Aural Warning—On the Ground

♦**Spoiler lever** not full forward
♦**Either throttle** advanced
Horn and warning ....***"SPOILER"***

Speed Brakes

Four flight spoiler panels move symmetrically
Max panel deflection is 35° from flush

Speed Brake Aural Warning—In Flight

♦**Flaps extended** beyond 6°
♦**Speed brake handle** not full forward
Horn and warning.....................................***"SPEED BRAKE"***

Spoiler/Flap Extended MC

Same conditions as above for Speed Brake Aural Warning Horn
De-energized in ground mode after landing with automatic spoilers armed until after retraction of flaps or speedbrakes

Ground Spoilers

Operate during landings and rejected takeoffs
Ground spoilers (Innermost panels each wing)
Flight spoiler panels (outer 2 panels each wing)
All 6 deployed 60° from flush

Auto Spoiler Operations on Takeoff

Arming—Squeeze lever; raise to armed position
Activation—Reverser thrust selected; Results:
Spoiler lever automatically extends ground spoilers fully (All panels, flight and ground, move 60°, see above)
Spoiler handle out of full forward activates auto-brakes after "short delay"
Delay defined as 1 second in **MAX** position, and
Approx. 3 seconds in **MIN** or **MED** ....LAND 20.6
Manual spoiler deployment also signals brakes

RTO Above 70 Knots—Max autobraking applied to full stop or pilot takeover
Uses pressure from both hydraulic systems
OM refers to this as "dual system braking"
(See also Vol. II, LANDING GEAR-20.5)

RTO Below 70 Knots—Minimum autobraking applied to full stop or pilot takeover
Uses single (right) hydraulic system (*Not in OM—from study tape*)

Auto Spoiler Operation on Landing

Arming: Spoiler lever raised
Note: Do not arm spoilers prior to gear extension (OM Vol. 1, LIM-10.19)

Activation:
Main Wheel spin up ***OR*** Nose Strut Compression—Flight spoilers deploy 60°
Above plus both throttles at idle and weight on main gear—Ground spoilers also deploy 60°
For either of the above,
♦Spoiler lever is moved full aft
♦Autobrakes activate if armed

Reversal of Activation
Left Throttle forward for go-around after spoiler deployment
♦Spoiler Lever Retracts
♦Autobrakes released

Warning: *If hydroplaning, auto spoiler deployment will not occur until ground shift at nose gear touchdown, but manual spoiler operation available at main wheel touchdown.* OM Vol. 2, FLT-C-10.9

Auto Spoilers—TWA Aircraft......TWA Supplement 45.1

No autobrakes
- Auto spoilers not available on takeoff
- Autospoilers may be armed for landing.................TWA

Flaps

Power Source—Both L & R Hydraulic systems

Dual actuators for each flap panel

Loss of either system—normal operations from the remaining system at a reduced rate

Flap Position Indicator

Dual needles for L & R flap

Separate transmitter on each outboard flap panel

Leading Edge Slats

General

Six separate panels on each wing

Mechanically connected to operate as a unit

Power from both L & R Hydraulic systems

Loss of either system—normal operations from the remaining system at a reduced rate

Airfoil Anti-Ice Protection provided

Slat Positions

Retracted

Mid-Sealed—Middle position, with trailing edge of slat panels still in contact with the wing leading edge
- Selection—Flap handle between 0/EXT and 13/EXT
- Movement of Slat—Hydraulic

Fully Extended—Slat full down/forward with panels separated from wing leading edge
- Selection—Flap/slat handle at or beyond 15/EXT
- Movement of Slat—Hydraulic/electrical by stall warning computers when handle moved beyond 15°

Autoslats

Provide positive stall prevention with flaps down

Activated through the stall protection system (*see below*)

Arming:
- Speed below 240 knots
- Flap/Slat Handle in 0 to 13 T/O (EXT) range
- Either of the two stall-warning computers detect approach to stall

Ground Self-Test
- ♦Takeoff Flaps (0-13) set
- ♦Slats extend from retracted to mid-sealed
- ♦Stall computers generate a signal to drive slats to Extend position
- ♦While slats extend beyond mid-sealed, DISAGREE and AUTO lights illuminate
- ♦Slats return to mid-sealed position
- ♦Slat TAKEOFF Light comes on

Invalid test indication:
- Auto Slat Fail Light on **MC**
- May be reset & test repeated

Indicator Lights

TAKEOFF—Flap/Slat handle in T/O range

DISAGREE—L and/or R wing slats disagree with each other and/or Flap Slat handle position

AUTO—Slats have been automatically extended from mid to extended by the stall warning system

LAND—Flap/Slat handle set at more than 26° and slats fully extended

Other Systems

Stall Protection System

Stall protection
- Provided from two independent electronic stall detection and indicating systems
- Inputs to each system
  - ♦Angle of attack vanes
  - ♦Horizontal stabilizer position transmitter
  - ♦Flap / slat position transmitter
- Either system provides three warnings and/or actions before fourth and actual "*STALL*" audible warning occurs

Stick pusher
- Fifth and final event
- Occurs only if:
  - ♦Both stall warning systems operational
  - ♦Both systems sense a stall condition
  - ♦Slats have been fully extended

Sequence of FIVE warning/activation: (**S**peed **ASAP**)
- **S**peed *Low* ....**SPD LOW** displayed on FMA pitch window
- **A**utoslats.....................Slats to full extend if mid-sealed
- **S**tick Shaker ....Activates approaching stall
- **A**ural warning ....Audible "**STALL**" & klaxon at stall speed; **STALL** lights illuminate
- **P**usher ....Activates if both systems sense stall **AND** slats fully extended

Stall Warning
- System provides stall ***warning*** with flaps/slats retracted
- Either stall warning computer can sense stall and signal stall ***warning***

Stall Prevention
- System provides positive stall ***prevention*** with flap/slat handle set in the 0/EXT to 13° range
- One stall warning computer sensing conditions approaching a stall activates stall ***prevention*** (Slat DISAGREE and AUTO lights ON)
- Both stall warning computers must sense stall to initiate ***Recovery*** (Stick Pusher)

Stick Pusher Activation results in:
- ♦Autopilot disengages if on
- ♦Stabilizer input is made to reduce G-force
- ♦Stick pusher push to inhibit Lights illuminate

Stick Pusher Deactivation
- ♦Stick shaker shuts off
- ♦Airplane experiences a reduction in G-Force
- ♦Accomplished by pushing either **STICK PUSHER PUSH TO INHIBIT** Light
- ♦May be overcome by physically pulling stick back

Stall Indication Failure Light
- ♦System Failure sensed ***or***
- ♦Disagreement between system 1 and 2 ***or***
- ♦Post stall pusher system is shut off by pressing either **STICK PUSHER PUSH TO INHIBIT** light or by "G-Switch" (reduction of G-forces)

Takeoff Warning

7 Items—Associate with items on Control Pedestal

Stab Trim Handle—1 item
- Stab trim not within ± 1 unit of the setting in the Long Trim Window
- Aural Warning ...."**STABILIZER**"

Spoiler Handle—2 Functions, 2 Warnings
- Spoiler lever not full forward
- Aural Warning ...."**SPOILER**"
- Auto-spoiler not armed with auto brakes armed
- Aural Warning:................................."**AUTO SPOILER**"

Flap/Slat Handle—2 Functions, 2 Warnings
- Flap Position not in agreement with Dialed Flap setting in window
- Aural Warning ...."**FULAP**"
- Slats not extended (either mid-sealed or fully extended)
- Aural Warning ...."**SLATS**"

Auto-Brake Panel—2 Switches, 2 Warnings
- Parking Brake Set
- Aural Warning ...."**BRAKES**"
- Auto-Brakes not Armed with Auto Spoilers Armed
- Aural Warning ...."**AUTO BRAKES**"

Alternate memory jogger: SAAB FSS
- **S**_tab Trim not Set for Takeoff_
- **A**_uto Brakes not Armed_
- **A**_uto-Spoilers not Armed_
- **B**_rakes_
- **F**_laps not Set for Takeoff_
- **S**_lats not Extended_
- **S**_poiler Lever not fully Retracted_

Flight Instruments

Note: *No attempt has been made here to completely describe the flight instrumentation displays and their connection to the FMS. These subjects are best learned using computer-aided color presentations, and are used on a day-to day basis by flight crews. Instead, key systems-related subjects have been summarized, which are easily forgotten and little used except for orals and recurrent training, and, of course, in the event of an actual malfunction. This section also serves as an overview for the transitioning crewmember.*

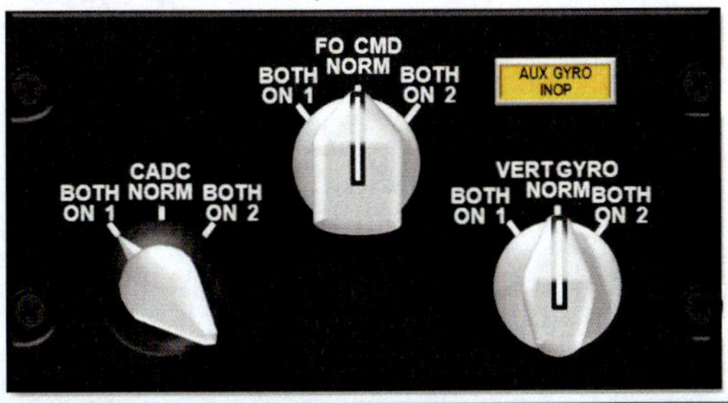

Primary Instruments

Primary Flight Instrument Categorization

Primary flight instruments
- ♦Mach/Airspeed Indicators
- ♦Altimeters
- ♦Vertical Speed / Resolution Advisory Instrument
- ♦ Attitude Indicators..................OM Vol. II, FLT-I 20.1-20.5

CADC Normal Operations

Sources for primary flight instrument information
CA From CADC 1
FO From CADC 2
Standby Instruments (center)—Alternate system comprised of auxiliary pitot/alternate static inputs

Altimeter Reporting Data

Transponder altitude data is independent of DFGS panel
Altitude data transmitted is based on 29.92 reference regardless of the baro setting on any cockpit altimeter
ATC Applies correction to altitudes below FL180

Static Air

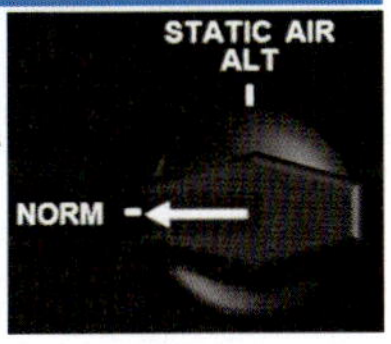

Static Air Source Switch
- Two possible sources for static *air*
- Can be used to select static source for CA and FO static instruments
- NORM—CADC Selector Switch selects static pressure source (see selector details following)
- ALT—Alternate pitot-static system selected
- May be selected on either side of the cockpit:

CADC Selector (Overhead panel)
- Three possible sources of Pitot-Static ***information***
  - ♦CADC 1 (CA system sensors)
  - ♦CADC 2 (FO system sensors)
  - ♦Alternate System

- With switch in **NORM**
  - CA Side CADC is #1
  - FO Side CADC is #2
- CADC BOTH ON 1; or BOTH ON 2
  - Places both sides of the cockpit on one CADC
  - Adjustments to altimeter setting may only be done from the selected side
  - Opposite side altimeter set knob is inop

CADC Light
- CADC Selector is out of NORM ***and*** Switching has occurred

Standby Airspeed Indicator & Altimeter
- Data always from Alternate Pitot-Static System
- Source cannot be switched

FD Light—FD Cmd Selector is out of the NORM position.

Radio Altimeters

Incorporated into EFIS displays
Rising runway symbol
- Activated at 200 feet
- Meets Center of instrument at 0 Feet

No ON/OFF Switch
Displays automatically when A/C ≤2,500' AGL
DH is green, followed by set value—DH 807
At DH—DH..XXX changes to yellow DH without altitude
Yellow DH symbol flashes for three seconds, then steady

Standby Instruments

Standby Attitude Indicator

Powered by DC Transfer Bus
Orange flag at "2 o'clock"
- Indicates power lost to indicator
- Should continue operating for approximately 7 minutes

Standby Altimeter

Pressure Operated Digital Counter
Near Zero Hangup
- Tends to hang up slightly passing 12 O'clock
- At "Zero" readings, pressure must be sufficient to rotate the cog moving the thousand foot counter
- Hang should not exceed a 25 foot lag or jump
- If it does, E-6 write-up is required

Static source is alternate static system

Standby Airspeed

Static source is alternate static system
Inputs are not corrected by a CADC

Standby Altimeter and Airspeed—TWA Aircraft

The two instruments are combined into one
See TWA Supplement....................OM Vol. II, p. 50.4 **TWA**

Other Instrument Systems

Overspeed Warning System—MAX SPD WARN TEST

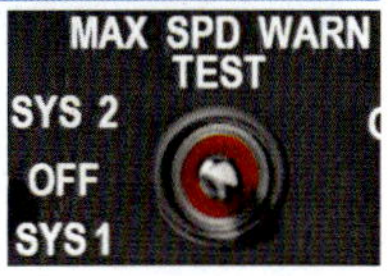

Dual System—1 & 2
- Both are tested on origination pre-flight
- System 1 Test—tests Captain's system
- System 2 Test—tests First Officer's system
- Aircraft Overspeed detection results in Clacker plus
- Aural Warning: ...."***OVERSPEED***"

Slat Overspeed
- Occurs if Slats not retracted at 280+ knots
- Aural Warning:................................"***SLAT OVERSPEED***"

Standby Magnetic Compass

Located above F/O seat
Designed to be out of the magnetic interference from electrical items in the forward panel area
Viewed with mirrors from pilot's stations

Instrument Error Tolerances

Specified for Primary & Standby Altimeter, Mach/AS, VVI, and Standby Attitude Indicator ....See OM Vol. 1 SYSTEMS-55.2

Attitude Indication Source

Vertical Gyros—4
#1-Captain, CB location F-15
#2-First Officer, CB location..F-2
Auxiliary, CB location...F-1
Standby ADI has dedicated gryo ....CB locationX-31
See CB diagrams, OM Volume II, ELECT-10.14, 10.17

Flight Management

This section contains only selected notes specific to the MD-80. Many aspects of navigation are common to other aircraft with which pilots will be familiar.

Performance Management System

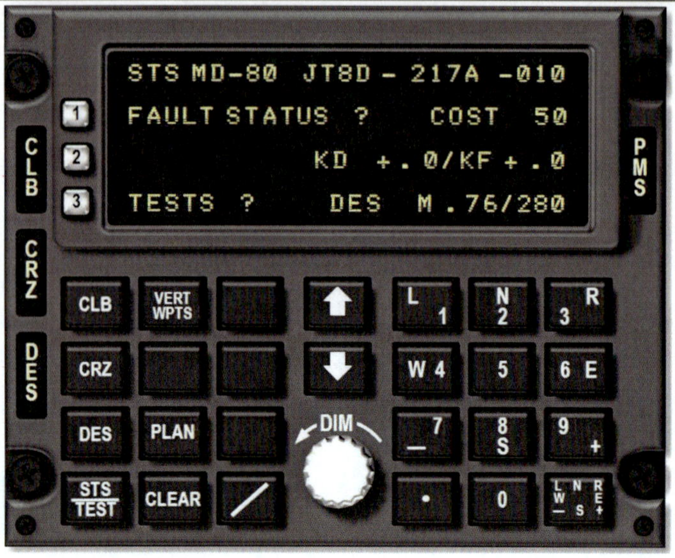

Fully integrated, selectable mode of the Digital Flight Guidance System (DFGS)

Computes a cost efficient flight profile based on
- Cost index value
  - A number from 0-255
  - Represents the relationship between fuel costs and the fixed costs of time related items
  - The lower the cost index. the more biased the computer is towards fuel savings
- Airplane performance, and
- Manual inputs.

Altitudes below 10,000 MSL—
- Automatic speed restriction of 250 knots applies
- Can be manually overridden

The PMS computed profile
- Includes a top of descent (TOD) point for idle thrust descent to bottom of descent (BOD
- Uses existing vertical speed and altitude (compared to the armed altitude) to determine its operating mode (CLB, CRZ, or DES).
- Each of these modes can be operated in
  - optimum (OPT) submode for minimum operating cost or
  - Non-optimum (NON-OPT) submode which uses a manually entered speed and / or rate of climb or descent.

Initial entry to the CLB, CRZ, or DES mode—
- PMS selects the submode (OPT or NON-OPT) which has been armed or selected
- If no submode armed or selected, the PMS automatically selects the optimum (OPT) mode for CLB and CRZ, and the non-optimum (NON-OPT) mode for decent.

Index is a factor of time vs. fuel prioritization
- If time is more important than fuel savings, cost index is higher
- If fuel savings are more important than time, cost index is lower

Performance Management System Lights on Pilot's Panels

CDU MESSAGE—Alerts to a message in the scratch pad area

VERTICAL ALERT Light—
- 15 Seconds to a PMS-generated vertical leg change (vertical waypoint)
- Comes on only when PERF mode is engaged
- Goes off when vertical leg change is completed or when CLEAR button is pressed

Multifunction Control Display Unit (MCDU)

Screen Format

Top line
- Title on left
- <Page number> / <of total on right>

Left & right sides
- Line-Select Key selectable fields
- Data can be selected to scratchpad, or inserted from scratchpad to the route information areas

Bottom line
- System generated messages
- Keyboard entries
- Data being moved to display field (scratchpad)

Fuel System

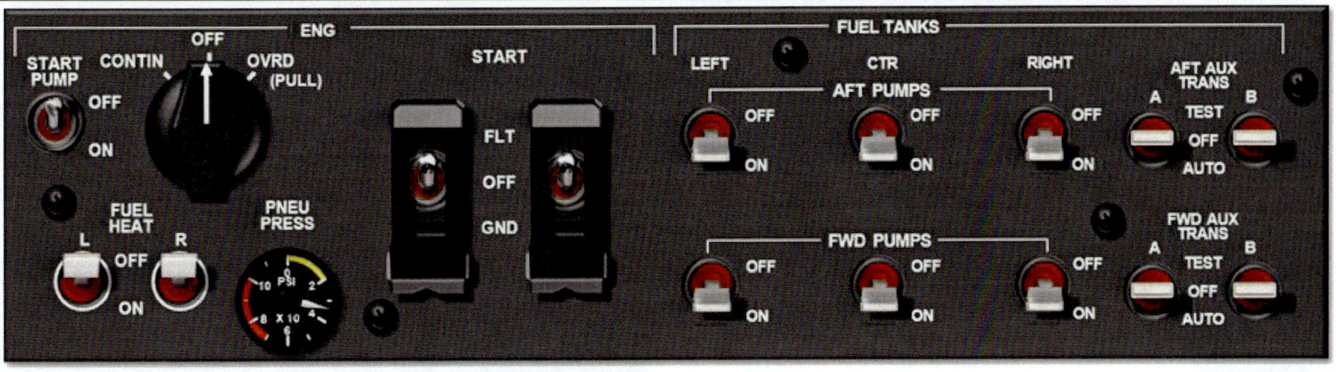

General

Fuel Tank Configuration

All aircraft—3 Tanks

1 Center Tank in Fuselage

Capacity ≈ 20,600 lbs.; no suction feed capability

2 Wing Tanks (Left Main and Right Main)

Capacity ≈ 9300 lbs.; suction feed capability

-83 Aircraft Only

2 Aux Tanks, 3785 Pounds Each

Total Fuel Capacity	Amount
Center Tank	20,596
Main Tanks (2 x 9,266)	18,532
Total, -82 Airplanes	39,128
Aux Tanks (2 x 3,785)	7,570
Total, -83 Airplanes	46,698

Fuel Feed & System Operations

Main Tank Boost Pumps

Each Tank has 2 boost pumps

-82 Airplanes have six total

-83 Aircraft have ten total

All 6 (or 10) pumps are identical in output pressure

Any one of the 6 (not aux) can supply both engines at takeoff power

Center Tank Pumps

Connected to each other in series

Allows center tank to feed fuel at a higher pressure, therefore before wing tanks if all boost pumps are operating

If one center tank pump fails—

Wing tanks tend to feed first

Center tanks have farther to pump fuel & more baffles, etc.

Fuel Transfer to Wing Tanks

With center tank pumps on and wing tanks not full, fuel may transfer to wing tanks

Maximum transfer rate—200 pounds per hour

Fuel Crossfeed Lever

Crossfeed OFF—Left wing tank feeds left engine and vice versa

Crossfeed ON—Allows left boost pumps to pump fuel to both left & right engine, and vice versa

Aux Tank Boost Pumps

Separate switches for each (Left & Right) pump in Forward and Aft Aux Tanks

4 Switches total

Aux Tank Fuel Transfer (Some Aircraft)

AUTO—Arms AC circuit to power applicable transfer pumps

Fuel transfer starts when center tank reaches approximately 13,000 lbs.

Note: Ensure fuel begins transferring by approximately 12,500 pounds center tank fuel

Total fuel center tank	20,596
Max center fuel used before transfer	-8,500
Difference	12,096

See OM Vol. I, LIM-10.24

Aux Fuel Pump Pressure Low (Fwd/Aft)

Comes on when pump is on but no fuel is being transferred

DC Start Pump

In Right Main tank

Used for APU starting when only battery power available

Powered by DC Transfer Bus

Power Sources

Sometimes referred to as "Lazy L" or "Lazy U"

L .... For Main Tanks (All tanks on -82s)

U .... For Aux Tanks included (-83s Only)

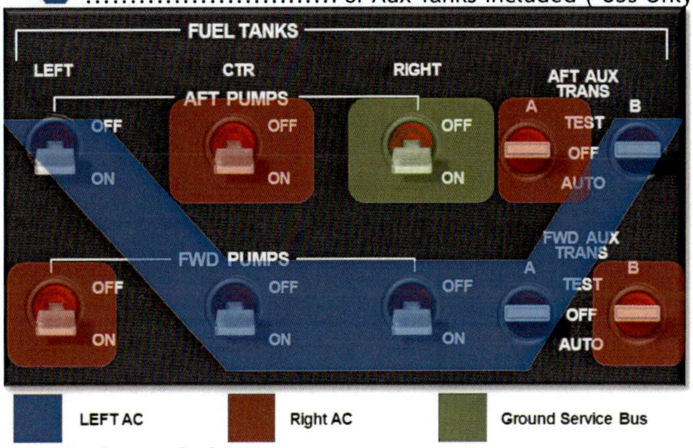

***-83 Airplanes Only**

****If powered,** ground service bus may be used for right aft boost pump to supply APU instead of DC start pump; (Dashed line is not on fuel panel)

Fuel Heat

Uses 13th Stage bleed air to heat fuel entering fuel filter

Use is for:

♦ Fuel temperature 0°C or below

Initiate on one engine, wait until engine parameters stable before starting other engine fuel heat

Momentary selection activates timer

Leave **ON** until a rise in fuel temperature and engine stable before selecting other engine

One cycle no later than 1 minute prior to T/O or approach

OM Vol. 1, SYS-65.1

♦ **FUEL FILTER PRESS DROP** Light ON **MC**

Blockage at filter is occurring

When Fuel heat is on, check

Fuel Temperature rise

Oil Temperature rise but below limiting temperature: Red Radial Line .... QRH ENG-16

Fuel temperature should be monitored for a drop when **FUEL HEAT ON** light extinguishes

Must be **OFF** for takeoff, approach, landing and go-around

Fuel Inlet Low Pressure Sensor

Indicates lower than optimal pressure at respective engine fuel pump

Normally indicates boost pump failure

Lights Illuminated:

L (or R) INLET PRESS LOW — L INLET FUEL PRESS LOW

MASTER CAUTION Light **MC**

See Vol. I Systems-65.1 for additional system operating considerations and parameters.

AUTO Position (Some TWA Aircraft)

Initiates 1 minute fuel heat cycle if

- ♦Fuel filter clogging is sensed *and*
- ♦In flight

If clog is still present after first cycle, a second minute is automatically generated

Procedurally, this switch is not used

See OM Vol. II, TWA Supp page 35.2 **TWA**

Fuel Loads

STD Load:

All fuel in wing tanks

If more fuel required than wing tanks can hold, remainder in center tank

After center tank full, Aux tanks fueled

***TO5 Fuel Load**

Used to hold nose gear down when towing or taxiing the airplane for maintenance purposes with insufficient fuel load for proper balance

5000 pounds in Center Tank

Remainder of required fuel in wing tanks (not necessarily full wing tanks)

010 Fuel Load

Fuel in center tank in hundreds of pounds, when wing tanks are not full

Range is 010-200 for 1,000-20,000 pounds

Example: 050 = 5,000 pounds

Remainder of fuel is in wing tanks

When used, max ZFW will be reduced by the amount of fuel in the center tank.

***ALT Fuel Load**

10,500 pounds of fuel divided between wing tanks

Remainder of required load in center tank

Used for Light passenger loads or ferry flights

SLP Fuel Load

Each wing is filled to 300lbs less than full due to steep ramp slope

Prevents fuel spillage

NAX—No aux tank fuel due to MEL restriction

NCF—Center tank empty or unusable due to MEL restriction

***NOTE:** *When fuel is in the Center or Aux tanks, and the Wing tanks are not full, the total of (actual zero fuel weight + Center tank + Auxiliary tank fuel) must be <122,000 pounds. See OM Vol. 1, p. LIMITATIONS 10.24, and Performance Manual LOADING p. 30.1*

Ballast Fuel procedures............OM, Vol. I, LIMITATIONS 10.24 *and* Performance Manual LOADING 30.1

Center Fuel Pressure Low

Low fuel pressure sensed in center tank with pumps on

Delay prevents false warnings from fuel sloshing........ **TWA**

Fuel Use

Feed Center tank fuel is used first

If Fuel in Aux tank, feed:

Center tank down to approximately 13,000 pounds remaining (but not below 12,000, see *note on previous page* and in the Limitations—Fuel section concerning no more than 8500 pounds used from center tank)

Then feed aux tank fuel into center tank

Then feed remaining center tank fuel

Than feed wing tanks, keeping them balanced

Fuel Quantity Displays

Separate Indicators for Each Tank

Test Display

Each Tank reads 3000 ±100 Pounds

Total is 9000 ±300

With Aux Tank, Total is 15,000 ±500

Gross Weight reads ZFW + Fuel

Aux Tank displays on Captain's Panel (Right, above)

Fault Displays

88888 = Normal Test display

Blank = Circuit Fault

Can Try alternate circuit (A vs. B)

99999 = Tank Fault at Probe

With bad probe, no alternate means to try

Dashes (- - - -) = component failure..................... **TWA**

Fuel Pounds Used Counters

Digital counters in each fuel flow indicator

Estimates fuel used based on fuel flow and time

Reset on pre-flights

Additional Fuel Notes

If a go-around is required with less than **1000 pounds of fuel** in either wing tank, avoid excessive or sustained nose up attitude in excess of 10°

Go around is not recommended with less than 500 pounds of fuel in each main tank. ....QRH FUEL 12.1

Hydraulics

General

Two independent systems
Each system consists of:
- Engine-driven hydraulic pump
- Reservoir
- Ground servicing point in respective wheel well

Electrically operated auxiliary pump supplements right system
Power transfer unit takes pressure from either system to power remaining system if it still has fluid

Reservoirs

In Main Gear Wells
Temperature sensor lights

- L (or R) HYD TEMP HIGH **MC**
- Activates at ≈200°F

Ram air is routed through each wheel well to cool reservoir and lines
Fluid quantity transmitter in each reservoir
- Quantity should be read with system pressurized
- Otherwise, air in the system can lead to false reading off-scale high
- If this indication is seen, make E-6 write-up

Engine-Driven Pumps

Selectable output
- 3000 psi.–HIGH–Upper Green Band
- ≈1500 psi.–LOW–Lower Green Band

Pressure sensor for warning lights

- L (or R) Hyd Press Low Light **MC**
- Activates at ≈900-1200 psi.
- Downstream of pressure gage sensors, so good gage pressure possible simultaneous with a warning light
- **Note**: May indicate the respective spoiler depressurization valve is closed (Instructor notes).

Auxiliary Pump

Electrically powered
Designed for continuous output at 3000 psi
Temperature protection—shuts down if overheat sensed
OVRD Switch position bypasses heat protection if output needed

Hydraulic Control Panel—TWA Aircraft

Pump switches labeled differently from AA aircraft
Switches are "functionally equivalent" .... **TWA**

Power Transfer Unit

Mechanically connects left & right sides
- Pressure is transferred from high side to low side
- High side acts as power source
- Low side acts as a pump

Unit control—single motor operating two shutoff valves
- Shutoff valve closes automatically if either system reservoir drops below a safe level

Unit located in Left main gear wheel well

Powered by Left System (LIE)

Left Thrust Reverser
Inboard Flight Spoilers
Elevator Boost

Powered by Right System (GRROS)

Gear
Right Thrust Reverser
Rudder
Outboard Spoilers
Stairs

Powered by Both Systems

Associate with key systems for ground stopping & directional control
- ♦**Flaps & Slats** (allows slower landing speed)
- ♦**Brakes**
- ♦**Nose wheel steering**
- ♦**Ground Spoilers**

Note: *Systems operate at reduced rate if one pump is powering systems on both sides (one pump inop)*

Backup Accumulators Provided for: (BEARR):

Brakes..2
Elevator Boost..1
Aft Stairs...1
Rudder (Shared with Aft Stairs)......................0
Reversers..2
Six Total..6

Items which assist with stopping the aircraft have two accumulators, one on each side. (Brakes & Reversers)

Electrical Power Failure

Engine-Driven pumps fail to the high output mode if the L & R DC buses are unpowered
Power transfer unit shutoff valves remain in the position they were in at electrical power failure

Spoiler Depressurization Valve

Used for maintenance tests
One in each wheel well for respective side system
On preflight—should be checked **ON** (pressurized)
- Makes a "T" with incoming hydraulic lines
- Three-position valve; **ON** (normal, pressurized for flight) is the "T" position

Power Sources

Respective L/R AC..........................Hydraulic pressure gage
....Hydraulic quantity Gage
....Brake pressure gage
Left AC...................Auxiliary pump for right hydraulic system
Right DC ....Power Transfer Unit control
....Right hydraulic pump control
Left DC.....................................Left hydraulic pump control

See OM Volume II, ELECTRICAL 13-19

ICE– Anti-Ice and Rain Protection

Anti-Ice

Anti-icing Definition

See Flight Manual Vol. 1, p. LIM 10.29 *or* *Page 12 of this Study Guide*.

Engine Heat

Each engine supplies its own heat
- No crossover capability
- Separate switches for each side

Ice protection for:
- Engine nose cowl
- Inlet bullet
- Compressor inlet guide vanes

Airfoil Anti-Ice

Engine bleed air feeds into common airfoil anti-ice duct
- Pneumatic crossfeeds open to allow air to airfoil
- Either engine can provide sufficient airflow

Heating air provided (in-flight only) for:
- Wing Leading Edge Slats
- Forward Strakes
- Air Conditioning Ram Air Scoop
- Horizontal Stabilizer Leading Edge
  - TAIL button diverts airfoil heating air from wing and forward strakes to horizontal stab
  - Remains diverted for 2½ minutes
  - Automatically reverts to normal mode
  - Can't stop 2½ minute timer once started
  - Use at three main times:
    - Every 20 minutes in icing conditions
    - One minute before extension of landing flaps
    - Prior to turning Airfoil Anti-Ice Off when icing conditions no longer encountered

Probe Heat—Electric

Heats the following:
- Pitot Tubes
- Static Port areas
- AOA Probes
- RAT Probe (Inhibited on ground)

Windshield Anti-Ice

Electric

Must be ON for all flight operations

For Limitations............ ...*See page 11 of this book*, *or* LIM 10.30

Overwing Heater System

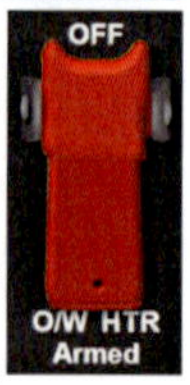

Caution: (Summarized from OM)
- ♦ ..*Ice shedding from upper wing surface can severely damage engines*
- ♦ ..*Clear ice forming over cold fuel tanks is main cause*
- ♦ ..*Ice forms most often on the inboard aft corner of main wing tanks*

Heating blanket covers approx. 45 square feet of upper wing surface in front of engine inlets

Rough ice-detection stripes painted on top of the blanket

On ground, single engine taxi without APU—system is load-shed

Operation
- Ground only—deactivated in flight by ground shift
- Maintains heated surface from 40°F (4°C)-85°F (29°C)
- Inhibited by **O/W HTR** switch placement to **OFF**
- Completely automatic in normal use

Warm Light—Indicates both heaters ≥ 40°F (4°C)

L FAIL or **R FAIL** Lights—any detected fault

Load-Shedding
- Occurs automatically with O/W Heaters powered
- Items shed vary power source available
- Sole Power Source............................Automatic Load Shed
- External *or* APU Only ....Galley 4
- Single Engine Generator......................All galleys, ***and both*** overwing heaters

OM Vol. I, SYSTEMS 35.1

Ice Detection Ring

Located on both L and R windshield wipers

Provides a representative surface to help determine whether ice may be forming on airplane surfaces

Warning Lights

Overwing Heater Warm/Fail Lights

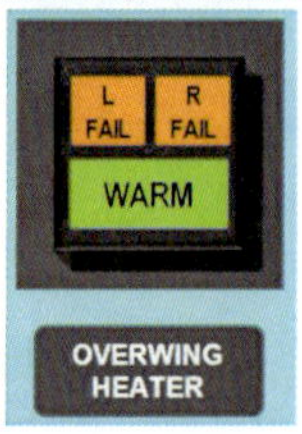

Located on Captain's instrument panel

L (or **R**) **FAIL** Light
- Fault detected with power applied

WARM Light
- Indicates power is applied ***and***
- Temperature of both heaters is over 40°F (4°C)

Note: *WARM and FAIL lights are inhibited when throttles are advanced for takeoff.*

L or R Engine Anti-Ice On

Indicates one or more engine anti-ice valves have fully opened

Note: *When any engine anti-ice valve is open, light will illuminate regardless of engine anti-ice switch position*

L or R Engine Valve Lights

Indicates disagreement between switch position and any engine valve(s) on the affected engine

Ice Protect Supply Press High Light MC

Excessive duct pressure (> 22 psi)

Malfunction of ice protection pressure regulating valve

Minimum duct pressure for anti-ice use—20 psi

Pitot/Stall Heater Off Light MC

Meter selector in **OFF** ***or***

Meter in **ON:**
- Electric power loss to one or more of the pitot tube or stall warning indicators

Note: *Does NOT come on for static port heater.power failure*

Airfoil Ice Protection Press Abnormal Light MC

Low or **unbalanced pressure** in wing and strakes ***or***
Low pressure in duct to horizontal stabilizer
or
With airfoil anti-ice switch in OFF—Malfunction of ice protection regulator
Note: If inadequate anti-icing available, add 10 knots to approach speed and use flaps 28° for landing.
....QRH ICE-1

Ice Protect Temperature Low

Air temperature in pneumatic crossfeed too low for operation of anti-ice
May be due to:
Low engine thrust ***OR***
Closed pneumatic crossfeed valve(s) ***OR***
Malfunction of the augmentation valve

Ice Protect Temp High MC

Air in pneumatic crossfeed is above normal operating temperature
Indicates malfunction of the augmentation valve

Operation

Note: *Most of the information in this section is from OM Volume 1, GENERAL, Cold Weather Operations 30.1 to 30.37. It is consolidated here for ease of study. See also* *pages 21-22 of this Study Guide.*

Engine Icing In Flight Indications

Ice buildup on probe may cause engines to indicate TRI power while actually developing less
If this occurs,
Throttles will retard
Depending on Autothrottle mode:
Airspeed may decay
Climb rate may decrease or reverse
Differing rates of ice buildup may cause disparity to be sensed by AT, which will then disengage in EPR LIM mode ...... OM Vol. 1 SYSTEMS 30.6

Engine Anti Ice Use

Continuous ignition ON with engine anti-ice in use

TWA
Continuous -**A** or -**B ON**

Use any time in icing conditions as defined in Limitations section
Turn on one side at a time
Allow engine to stabilize before turning on other side
Ground Use
Turn on after each engine is stable after start
Do not stabilize between 61-74% N_1 (fan blade damage)
For "significant precipitation" see below
(See OM Vol. 1 SYSTEMS 75.1, 75.2)
If in light or minimal icing conditions—Engine ignition selector **OFF** ....OM Vol. 1, GENERAL 30.4, (Ground) ***and*** OM Vol. 1, GENERAL 30.7, (In Flight)

Airfoil Anti-Ice Use

Use any time in icing conditions as defined in Limitations Section ***except:***
Do not use for takeoff, and then turn on at 1000' AFL
When used, Engine Anti-Ice must also be on
Maintain sufficient thrust to keep **ICE PROTECT TEMP LOW** lights out
Pneumatic crossfeed valves open when in use
Minimum Duct pressure 20 psi
After turning on, L or **R ICE PROTECT TEMP LOW** light(s) may be on for up to 1 minute

One or both ICE PROT TEMP HIGH light may come on if pneumatic crossfeed levers left open after termination of anti-ice use

WING ANTI-ICE LIGHT indicates anti-ice heat has been selected for wing leading edge and forward strakes.
Tail De-Ice

When button is pushed
Air is diverted to tail for 2 ½ minutes
Automatically reconfigured to deice wing leading edges after this perio

TAIL DE-ICE ON light indicates tail heat has been selected

Exceptional Icing Condition Measures

Ground Use with "**significant precipitation:**"
Run up engines no more frequently than once every 10 minutes to
As high a thrust as practical for
70% N_1 for a minimum of 15 seconds is desired, or
Alternately 60% N_1 for a minimum of 40 seconds
No more frequently than every 10 minutes
Subsequent takeoff under these conditions
Should be preceded by a static run-up to as high a thrust level as practical *with*
Observation of EPR and EGT to assure normal engine operation
Takeoff in known moderate icing conditions
At pilot discretion use ignition **OVRD**
Select **CONTIN** as soon as practical after climb configuration established
In-Flight use with severe icing conditions
Minimum desired N_1 70%
Thrust reductions below 70% should be
♦No lower than 55% ***and***
♦Limited to 1 minute
Engines should be run back up to 75% N_1 for at least 1 minute following reduction below 70%....OM Vol. 1, GENERAL 30.7
APU Recommended on Contaminated Runway Landings
APU should be on prior to final approach
Provides backup electrical power source if generator(s) are lost due to slush or water ingestion
OM Vol. 1, GENERAL 30.10

After Landing

Leave flaps/slats extended to at least 15 / EXT if
Making approach in icing conditions ***OR***
Landing with snow, slush or ice on landing runway
Allows a check of flight controls for damage, and makes deicing easier
Check may be accomplished by deicing personnel or flight crew member ....OM Vol. 1, GENERAL 30.11

Turbulence Considerations

Use speedbrakes to slow airplane
Thrust
Use smooth power changes
Maintain thrust as high as possible
Once set to maintain speed, avoid further changes
If throttles are in idle when heavy rain encountered. Monitor N_2, as its decay may be the first indication of a spool-down
Don't "chase" speed or altitude
Other Considerations.............................QRH ICE 17-18

Landing Gear

Landing Gear

General

Tricycle Landing Gear

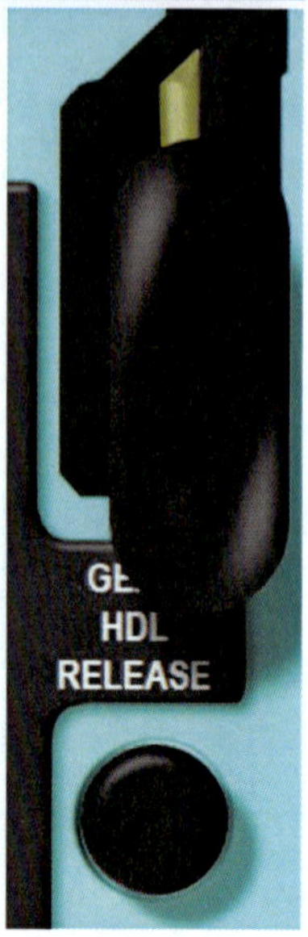

Mechanically controlled—gear handle connected by cable and linkage to transfer valves

Hydraulically operated
- Pressure from right system *OR*
- Pressure from left system transferred to right system by Power Transfer Unit

Backup Operation with complete hydraulic failure
- Mechanical handle to release uplocks
- Gear free-falls
- Locks down with over-center locks

Spray deflectors on all three gear

Ground Shift Mechanism

Located on Nose wheel

When NOT compressed:
- Disengages NW Steering
- Centers Nose wheel for retraction
- Retracts landing gear handle release button

Activates 2 switches—Switches establish ground or flight mode for other systems

Chart on page LANDING GEAR 10.18 of OM Vol. 2 shows all items affected by ground shift mechanism

Nose wheel Steering

Mechanically Controlled

Hydraulically Actuated

Bypass ValveManually Operated
- Used to disconnect hydraulics from steering actuators for towing

Steering available
- Through rudder pedals—17° L or R
- Through steering wheel—82° L or R
- If Gear Lowered by the alternate system—
  - Right system pressure blocked
  - Less than normal authority to the left

Gear Locking

Main gear locked down with over center linkage

Main gear locking in **UP** position
- Hydraulic pump output **high**, pressure holds gear up
- With pressure **low**, gear rests on main gear doors, which are locked up by uplocks

Nose gear locking—Over center locks up and down
(See OM Volume II LANDING GEAR 10.11)

Gear Doors

Main Gear
- Main Gear doors hydraulically actuated
- Outboard Gear doors mechanically linked to gear movement

Nose Gear has 4 doors
- 2 Aft doors are linked mechanically to gear
- 2 Forward doors can be opened manually on ground for work in nose wheel bay

Landing Gear Status Lights

Position Lights

- <u>Green</u>Gear handle down and associated gear in down-and-locked position
- <u>Red</u>♦Any unsafe gear condition *or*
  - ♦Landing gear in transition *or*
  - ♦Landing gear not in agreement with gear handle *or*
  - ♦Gear handle down & any gear not locked down *or*
  - ♦Gear up & locked & either or both throttles closed

GEAR DOOR OPEN Light On any time main gear doors are not locked closed

Landing Gear Warning Horn and Aural Warning

Warning Horn
- Sounds when
  - Below 1000' AGL on Captain's RA* *AND*
  - Airplane below 210 Knots *AND*
  - One or both throttles closed *AND*
  - Gear not down and locked
  - * If Captain's RA inop, sounds at any altitude with above conditions
- May be silenced by pedestal-mounted button unless
  - Gear not down & locked ***and*** flaps extended over 26°

Aural/Vocal Warning
- Landing Gear not down and locked
- Flaps extended beyond 26°
- May ***not be silenced*** until gear is lowered or flaps < 26°
- Aural Warning ...."LANDING GEAR"

Gear Down Verification with Power Loss

Main Gear
- Periscope between 3rd & 4th windows aft of aft overwing exit
- Cover may only be removed when cabin is depressurized

Nose Gear—Pedestal-mounted pin extends when nose gear is locked down

Brakes

General

Brake actuators on **each main gear**

Powered by both hydraulic systems

Components
- 4 Disc pistons on each wheel from each hydraulic system
- 8 Power pistons per wheel
- Accumulator on each main gear
  - Sufficient for approximately 5 brake applications
  - Brake pressure indicator monitors hydraulic pressure on each system
- Brake temperature gage

Parking Brakes

Effective if brake pressure gage is above the RED arc

Aural warning if ON and throttles advanced for takeoff

PARKING BRAKES ON Light

- Parking brake set
- If on with brakes not set, indicates antiskid malfunction

Wheel Not Turning Light

Speed difference between **fastest** and **slowest** main gear wheel exceeds 20%

Anti-Skid

Armed by switch on overhead panel
- **ARM**—After main wheel spin up, allows anti-skid system to modulate brake pressure to prevent skidding
- **OFF**—Disarms Anti-skid & Acts as a reset position

Circuit caution lights

- Separate lights for each circuit (**INBD** & **OUTBD**) on each side (**L** & **R**)—4 lights
- Indicate one of the following:
  - Respective circuit failure is detected
  - Anti-skid arming switch is off
  - Test switch in TEST, if gear are down

Deactivated by (LAPS):

Landing gear handle not in down detent
Arming switch OFF
Parking brakes set
Slow taxi speed (Below approximately 10 knots)

Brake Temperature

Gage and warning light indicate temperature

Excessive temperature may result in tire fuse plug melting & loss of tire pressure

Key Temperatures

Max temperature for takeoff ....205°C

Overheat light ....ON at 305°C, OFF at 260°C

Hot Brakes ....200-400°C

Overheated Brakes ....>400°C

Maximum to set Parking Brake ....300°C

Brake Temperature Test Button ....425-475°C ....and **OVHT** Light ON with button depressed

Brake Wear Indicator Pins

Limits, with ***brakes parked***—pins must extend beyond brake housing as follows:

MD-82s—Pin flush or greater

MD 83s—Pin extended more than ½" above flush

See OM Volume 1, PRE-FLIGHT 10.14 and LIM 10.31

Automatic Brake System

ABS Takeoff Mode Arming (FAAST)

Flaps positioned to less than 26°

ARM-DISARM Switch to **ARM**

Anti-Skid **ARM-DISARM** Switch to **ARM**

Spoilers Stowed

T. O. Position on **AUTO BRAKE** selector

Take-off warning sounds if ABS not armed with spoilers armed, and vice versa

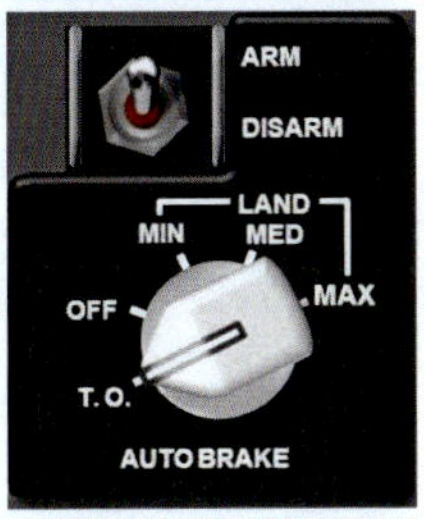

ABS Landing Mode Arming (AAA)

Anti-Skid Armed

ABS Switch in a LDG range (Not OFF)

ABS Arming switch to ARM

In addition to the above conditions, the landing gear handle must be down to arm ABS for landing

ABS Activation on Landing

Throttles retarded below 22°

Brake pedals NOT depressed

Spoilers deployed (manually or automatically—*see page 38*)

Automatic delays following spoiler deployment (allows for nose wheel touchdown):

1 Seconds in MAX

3 Seconds in MED or MIN

ABS Deactivation

Brake system reverts to manual braking if any of the following occur (**F TABS**):

F*laps raised to less than 26° with speed above 70 knots*

(Below 70 knots, raising flaps to less than 26° does ***not*** *discontinue ABS)*

T*hrottle—Either advanced beyond 22° above idle*

A*RM-DISARM Switch to DISARM*

B*rakes—Either brake pedal depressed more than 25%*

S*towing Ground Spoilers*

Releases brake pressure without disarming ABS

Re-deploying spoilers will reactivate ABS until it is disarmed by any other means

At ABS disarming, the following occur:

ARM-DISARM Switch drops to DISARM

ABS Lights come ON

Rejected Takeoff (TO) and ABS

Rejected takeoff is signaled to ABS by spoiler handle moving aft manually or automatically

Takeoff rejected below 70 Knots

ABS reverts to landing mode

MIN braking force applied

Rejected takeoff above 70 Knots

MAX Autobraking

T.O. Is the ***only auto mode*** which uses ***both*** hydraulic systems

Other modes (**LAND MIN**, **MED**, and **MAX**) use ***only*** right system .See schematic, OM Vol. II p. LANDING GEAR-10.17

Pilot takeover initiated by:

Throttle(s) forward

Brake pedal(s) depressed

ABS Fault Detected

Causes the system to **DISARM**

Results:

ARM-DISARM Switch drops to **DISARM**

AUTO BRAKE FAIL Light **MC**

ABS Lights come **ON**

Rearming Autobrake following fault detection

AUTO BRAKE Selector switch to **OFF**, then to desired deceleration setting

ARM-DISARM Switch to **ARM**

Autobraking will resume if the fault has cleared

Preflight Tire Checks

Tire Inspection / Replacement Guide gives guidance for tread wear considerations.............OM Vol. 1, SYSTEMS 80.1-80.3

Miscellaneous Systems

Lighting

Strobe Lights

Forward and aft wing strobes mounted at wing tip
ON When:
- POS/STROBE Switch in BOTH
- Nose wheel off the ground

Landing Lights

Nose Lights
- Two bulbs
- ON When landing gear handle is DOWN with switch on

Wing Tip Retractable Landing Lights
- Separate switch positions for **EXT OFF** and **EXT ON**
- Extend when Switch in **EXT** (**EXT OFF** or **EXT ON**) ***and***
  - Both engines running ***or***
  - One engine running with gear extended
    - On Single-Engine Go-around, lights turn off and retract
    - Design allows for automatic retraction in the event of engine failure and subsequent go-around
  - Note: Control may be regained after engine failure logic activates by cycling switch to **RET** and back to **EXT**

Potable Water Systems

Water Shutoffs

See OM Volume II, MISCELLANEOUS 60.1, 60.4 for location
Valves can be turned past CLOSED position, so use care in turning OFF to ensure proper position is not passed
Water flow ceases approximately 30 seconds after valve is turned OFF

Water Servicing Point

Guarded switches should be closed on pre-flight
To get water pressure to lavatories and galleys:
- Fill & Vent Valve Switch (Some airplanes)—Must be in CLOSE (Guard Closed)
- Water quantity Mode Selector (Some airplanes)—Must be in OFF
- One pack must be ON for pressure

Passenger Information Signs

NO SMOKING (NS) Signs & Switch

Switch Positions—all deactivated
Placard—installed over light (does not illuminate)

FASTEN SEAT BELT (FSB) Signs

Switch activates **FASTEN SEAT BELT** and Lavatory **RETURN TO CABIN** Lights
AUTO—ON when Slats are extended
ON/OFF—Self Explanatory
Note: *Cabin over 10,000 Feet—FASTEN SEAT BELT signs come ON automatically*

Flight & Cockpit Voice Recorders

Digital Cockpit Voice Recorder

Stores 2 hours of voice data
Uses solid state technology rather than tape
Airplanes manufactured after October, 1991
- Require boom mics
- Placard installed on both sides of instrument panel:

BOOM MIC REQUIRED BELOW FL180 ON THIS AIRPLANE

TEST Button
- Pressed for 5 seconds
- Tests all channels

Test monitor meter
- Needle does not move until all channels have been tested
- Needle then goes into green band if test is good, remains there until button released

ERASE Button—Tape erased in these conditions:
- ♦Button held for two seconds .... ♦Parking brake set
- ♦Airplane on ground .... ♦AC Power available
- Note: Do not hold longer than 5 seconds

Flight Data Recorder

NORM—Flight Data Recorder runs when
- Parking brake is released and
- Either fuel lever is ON

GND TEST—Interlocks (above) are bypassed
FLIGHT RECORDER OFF Status Light
- Recorder is de-energized
- Tape is broken, exhausted, or not winding properly

Cockpit Oxygen System

Oxygen Supply

Single high-pressure gaseous cylinder
Located in cockpit behind F/O seat
Pressure—1100 psi at 70°F
Shutoff on bottle
Thermal discharge disc outside airplane, right side, indicates bottle expended

System Use

Diluter Demand Control
- 100% OXYGEN—provides 100% O_2 at all altitudes
- NORMAL—
  - O_2 is mixed with ambient air to maintain acceptable volume of O_2 to the pilot for the current altitude
  - Above 28,000'MSL, O_2 is delivered under pressure

Emergency O_2 Switch
- EMERGENCY—
  - Provides O_2 under pressure to pilot's mask
  - Pin must be pulled to place switch in this position
  - Feature provides a backup means of getting high pressure O_2 in the event of a failure of the normal system above 28,000'
- NORMAL—Diluter Demand Control switch determines O_2 flow rate
- TEST MASK—
  - Supplies O_2 under pressure for checking mask and hose assemblies
  - Spring-loaded back to NORMAL position

Flight Attendant & Cabin Oxygen Systems

Flight Attendant Portable O2 Cylinders

Four bottles located in cabin
Pressure—1800 psi
Normal flow—30 minutes
Flow Rate—4 Liters per minute

Passenger Oxygen

May be provided by AA with prior coordination for passengers with medical needs
Passengers may NOT provide their own oxygen
Security concerns—canister could contain anything...

Passenger Oxygen System

Components

Chemical O_2 Generators
- One at each position (See below)
- Generate O_2 for at least 15 minutes
- Generation rates, and therefore flow rates are low
- When activated, an odor results which can be disconcerting if not expected

O_2 Reservoir bags
- Collect generated O_2 between breathing cycles
- Bag will likely not inflate, depending on:
  - Cabin altitude
  - Time since generator activated

Inflation indicator
- Small green bag at inlet end of mask reservoir bag
- Inflates slightly when O_2 is flowing

Individual Unit Locations:

Each Passenger Row
- Three-seat rows (right side) have four masks
- Two-seat rows (left side) have three masks

Forward And Aft Flight Attendant Stations (Mid Attendant seat gets O_2 from one of the passenger-type overhead masks)

Lavatories—None (All removed)

Activation

Automatic—Cabin above **14,000'** pressure altitude
- Doors open
- Any one mask being pulled down to separate lanyard from O_2 generator starts O_2 flow from the generator to all masks in the unit

Manual Override
- **EJECT**—Opens all O_2 compartment doors
- Hold switch 3-5 seconds, but not over 5 seconds

Power
- Left AC & DC normally power door activation
- If power loss occurs, right AC & DC provide power

Cabin Oxygen On Light

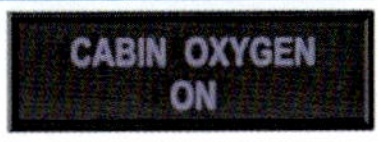

Indicates
- **Electrical power has been applied** to O_2 compartments door solenoid

Emergency Equipment Location

16 First Class, 24 Coach

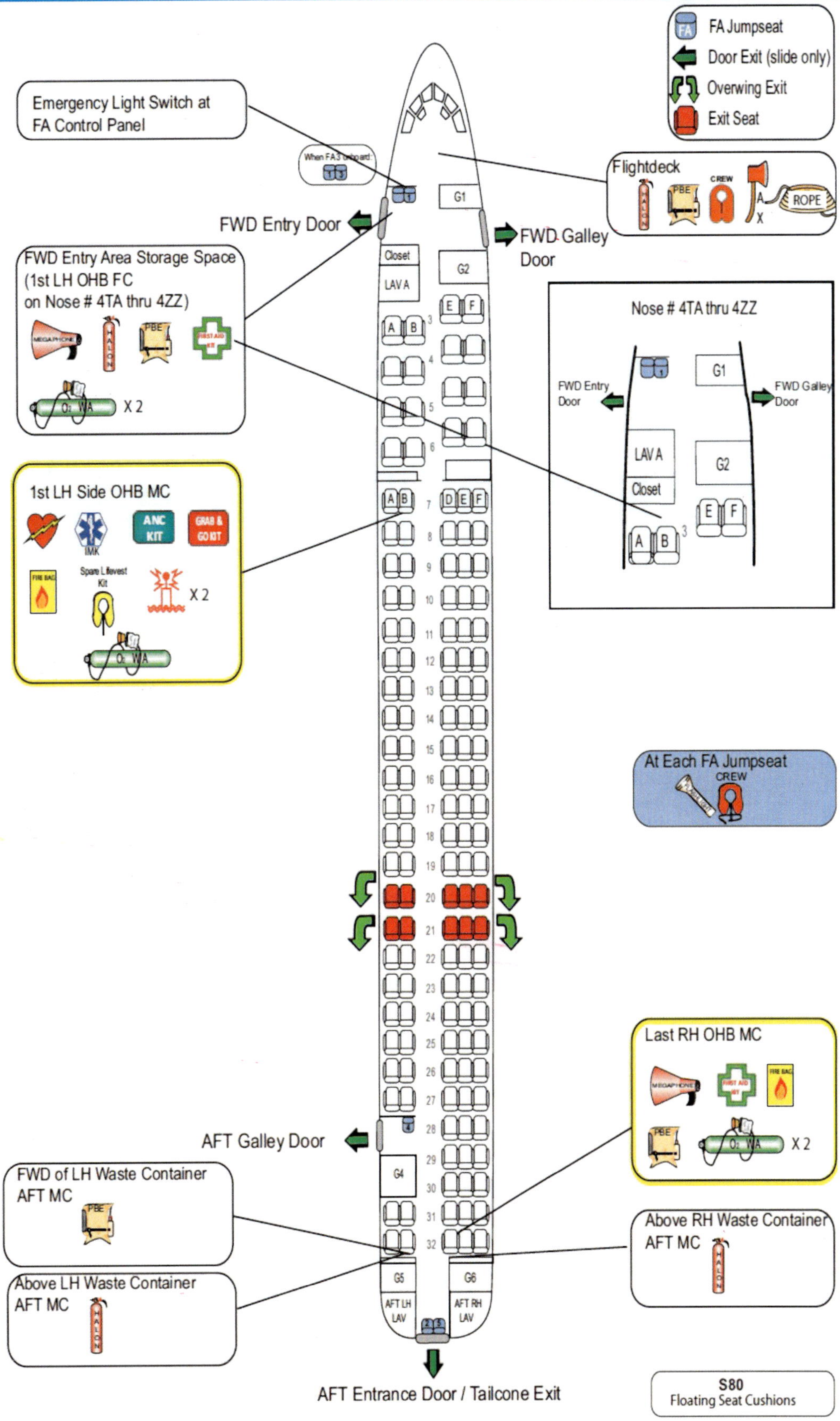

Reference—OM Volume 1, GENERAL 15.6

Navigation

Compass Systems

Two independent systems
- Each system stabilized by an associated directional gyro
- Magnetic inputs from an associated flux valve
- Displayed at all times except when RAD / NAV switch is in NAV

Compass system 1 provides inputs to
- FO compass indicator
- Captain's ND, VOR / LOC 1
- Digital Flight Guidance System (DFGC) 1 & 2

Compass system 2 provides inputs to
- CA compass indicator
- FO's ND, VOR / LOC 1
- Digital Flight Guidance System (DFGC) 1 & 2

VHF Navigation Systems

Two independent systems
- VOR / LOC 1, controlled by CA VHF Nav Control Panel
- VOR / LOC 2, controlled by FO VHF Nav Control Panel

VOR / LOC 1 provides inputs to
- CA ND and PFD
- CA and FO Compass indicators
- DFGCs 1 and 2

VOR / LOC 2 provides inputs to
- FO ND and PFD
- CA and FO Compass indicators
- DFGCs 1 and 2

Automatic Direction Finding (ADF) Systems

Single ADF

Control panel on forward pedestal
- Selects operating mode and frequency
- Displays on both CA and FO **VOR / ADF 2** pointers

Displays
- Two VOR / ADF selectors on each compass indicator
- Only the right ADF selection position functions

Marker Beacon System

Pre-tuned

Provides visual and aural signals to three dimmable lights on CA and FO instrument panel

Warning & Alert

Warning & Caution Lights

MASTER CAUTION Light

Illuminates for activation of certain individual caution lights on annunciator panel
Pressing either light extinguishes both **MASTER CAUTION** Lights and reset the system for subsequent indications.

MASTER WARNING Light

Illuminates for activation of certain individual caution lights on annunciator panel
Pressing either light extinguishes both **MASTER WARNING** Lights and reset the system for subsequent indications.

GPWS

Ground Proximity Warning System

GPWS Warning Categories GPWS Warnings can be divided into 5 general categories:
- ♦Excessive Sink Rate
- ♦Excessive Terrain Closure
- ♦Altitude Loss after Takeoff
- ♦Descent in Wrong Configuration
- ♦Descent below Glideslope

GPWS Mode Summary

Gives warnings for 5 modes/terrain closure situations

<u>Excessive Descent Rate, 50-2450 Feet AGL</u>
Ⓢ*Repeated every 0.75 sec.* .... ***Sink Rate, Sink Rate***
Ⓗ*Continuous* .... ***Whoop-Whoop Pull-Up***

<u>Excessive Terrain Closure, 0-2450 Feet AGL</u>
≈30 Seconds to Ground @2500'
≈20 Seconds to Ground @1000'
≈10 Seconds to Ground @500'
Ⓢ*Rapid Succession* .... ***Terrain-Terrain***
Ⓗ*Repeated every 0.75 sec* .... ***Whoop-Whoop Pull-Up***

<u>Altitude Loss after T/O</u> (65-700 Feet on takeoff; or following gear or flap retraction <200 feet on Go-around)
Repeated every 0.75 sec .... ***Don't Sink***

<u>Terrain Clearance</u>, in Wrong Configuration
<u>Gear Up Below 500'</u>
<.35M- Repeated every 0.75 sec .... ***Too Low Gear***
>.35M- Repeated every 0.75 sec. .... ***Too Low Terrain***
<u>Gear Down, Flaps not in Landing Position Below 1000'</u>
<.29M, 50-200'-Repeated every 0.75 sec .... ***Too Low Flap***
>.29M, 50-1000'-Repeated every .75 sec. ... ***Too Low Terrain***

<u>Below Glideslope [?] 1.3 Dots <1000' RA</u>
Aural Warning (Repeated) .... ***Glideslope***

Function Ⓢ-Soft Warning Ⓗ-Hard Warning

<u>Notes</u>:
- ♦*Above examples and numbers are "ballpark only," intended for familiarization.*
- ♦*See Charts, pp. WARNING & ALERT 20.3 - 20.8 for details*
- ♦*For all of the above, **GPWS** and **GLIDESLOPE** Warning Lights (as appropriate) illuminate*
- ♦*Hard Warnings* Ⓗ *occur at performance approximately 10-15% more severe than Soft Warnings* Ⓢ *in Mode 1, and 20-40% in Mode 2.*

GPWS Fail Light

Self Explanatory

CADC Inputs

CADC 1 provides barometric sink rate
No switch to CADC 2 is possible

Enhanced GPWS Features (EGPWS)

General

Major features:
Worldwide database with all runways over 3500' long
(Not all of these can be displayed in map mode)
Terrain data from worldwide database can be shown
Warning capability if airplane flight path predicted to intersect terrain within specified criteria

Does not override non-enhanced GPWS—same warnings based on radio altimeter and descent rate combined with airplane configuration still in effect

Uses GFMS to compare airplane position to terrain database

Alerts

Caution Alert
Criteria—40-60 seconds from predicted terrain conflict
Conflict criteria
Within about 1/8 mile either side aircraft
The initial distance of 1/8 mile (1/4 mile wide) expands ±3° from both sides of aircraft as the path is projected forward
Aural alert, amber **TERR** annunciation on
ND (EFIS airplanes) or
Weather radar indicator (non-EFIS airplanes)
Illuminated **GPWS** light

Warning Alert
Criteria—20-30 seconds from predicted terrain conflict
Alerts—same as caution, except red **TERR** annunciation

Terrain Clearance Floor (TCF) feature
For all runways over 3500 feet long, specific terrain alerting
Creates terrain clearance envelope directly related to distance from the runway

Approach and departure
Criteria modified to avoid nuisance alerts

EGPWS Uses FMC data for proximity calculations
Inaccurate FMC position data can compromise EGPWS warnings
Man-made obstructions are ***not*** programmed

Display

Variable density dot patterns (All altitudes with respect to airplane altitude)

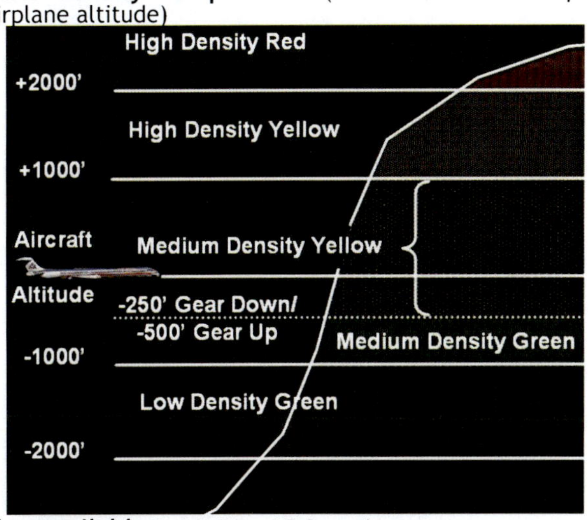

Display available in MAP or ARC modes:
Green **TERR** message annunciated when active

Activating display
EGPWS uses same line to HSI as weather radar
Weather radar and **TERR** must be **on** to display terrain data
Terrain and weather ***cannot*** be simultaneously displayed on the same HSI (*Can* be displayed on opposite sides of the cockpit, however)
Automatic activation—potential conflict will activate EGPWS display and appropriate warning ***if***:
Weather radar is **ON**
Terrain System Override Switch not in **OVRD**

Traffic Alert & Collision Avoidance System (TCAS)

General

Display
- TCAS display on dedicated TCAS Indicator
- Radar on Navigation Displays (ND)

♦**Resolution Advisory** (TA/RA) Mode may be selected as close to the departure runway as practical

♦**May be left in TA/RA** until after landing

TCAS Operating Policy

Conflict between RA and **ATC**—follow RA
- Return to ATC clearance after ***CLEAR OF CONFLICT***
- Advise ATC of deviations

Once RA issued, maintain speed as computations based on speed at the time

High-speed buffet—relax pitch as necessary to reduce buffet, but continue the maneuver

Visual contact made with traffic—Maneuver as required is allowed, however
- If following the TCAS direction will accomplish deconfliction, it should be followed
- Visual contact could be with incorrect traffic

Responding to RA, maneuver only as much as needed to satisfy the RA

Other critical warnings take precedence over TCAS
- Windshear
- GPWS
- Stick shaker or initial buffet—accomplish stall recovery procedure ....See QRH MANEUVERS 10

Altitude Band Selection

ABOVE—
- Used in Climb
- Traffic displayed -2700 to +8700 from present altitude

NORM—
- Used in cruise
- Traffic ±2700' from present altitude

BELOW—
- Used in Descent
- Traffic +2700 to -8700' from present altitude

Traffic Symbols and Warnings

Other Traffic .... White Open Diamond
- **More than 6 miles away *OR***
- Over 1200' altitude differential

Proximate Traffic......................... White Filled Diamond
- **Within 6 Miles *AND***
- **Within ± 1200'** altitude

Traffic Advisory (TA) Intruder .... Amber Filled Circle
- A target whose altitude is projected to be within ±900' at point of closest passage
- TA Alert occurs at 40 seconds from point of projected closest passage
- Aural Warning .... "***Traffic Traffic***"

Resolution Advisory .... Red Filled Square
- A target whose altitude is projected to be within ±600' at point of closest passage
- Threat is 25 seconds from closest point of approach
- Vertical avoidance maneuver displayed on IVSI
- *Aural Warnings* ....See OM Vol. II, WARNING & ALERT 60.15

TCAS II System

Installed on all aircraft

Requires flight crew response
- Within 5 seconds of initial RA
- Applying G-forces of ±.25 G
- This is more than AP can supply, so response must be hand-flown (*no change from previous TCAS policy*)

RADAR

Antenna Control

Stabilized in Pitch and Roll

Inputs to stabilization—
- Normal—VG 2
- VERT GYRO Switch in R on AUX—Aux Gyro

Controllable range
- Tilt: ±15° Pitch
- Range: 10, 20, 40, 80, 160 & 320 NM

Color Codes

Red—High density precipitation

Yellow—Medium intensity precipitation

Green—Low density precipitation

Black—Precipitation levels below pre-set threshold

Fault Monitoring

Automatic checks occur for system faults

Fault codes are displayed on Radar Indicator

Line Replaceable Unit (LRU) faults displayed only when mode selector is in **TEST**

Spoking
- Caused by external interference
- Not a system fault unless it continues for a long period of time

Performance

Editor's Note: Very few Performance Section items are expected to be known during orals and review sessions. But, (you guessed it!) there are a few. A few relatively common ones are summarized here.

TPS Computations

All TPS basic performance numbers are for dry conditions
Must apply contamination correction for wet runways

Standard Takeoff Thrust Not Authorized

Weather 4
- **Tailwind**
- **Contamination**: Contaminated Runway
  - Standing Water
  - Slush
  - Snow (wet or dry)
  - Ice
- **Windshear** reported or expected
- **Temperature** (Actual) hotter than Max Temp Standard (MTS)

Weight 2
- **Closeout**—Actual TOW from closeout is greater than Assumed Takeoff Weight
- **MEL/CDL Items** which specify MAX thrust required:
  - MAX-PNLT appears in the assumed temperature column
  - Conditions where weight correction is required and takeoff data is not automatically corrected by TPS

Airport 1
- **Mandated**—Flight Manual Part II Airport Advisory specifies use of MAX Thrust

See Performance Manual, TPS-10.4

Speed Assumptions

Built into flight planning computer algorithms

MAX	STD	MIN
.79	.76	.66

Follow fuel conservation cost index considerations in flight plan
Initial planning is done at STD Mach
If block to block is less than scheduled, lower Mach number plan is generated
Iteration continues until LRC "Floor" is reached or block time equals schedule
Final time is designed to represent lowest cost option, including crew costs and other known and programmed variables

Cruise Considerations

If temperature is warmer than planned
TAS increases 1 knot for each degree *warmer*
If EPR gage is inoperative
Match N_1 to engine with operative EPR gage
Below 320 knot/mach crossover
320 Knot cruise is computed

MAX Takeoff Weight (MTOW) codes (LASTED)

L=*Landing Weight Limit*
A=*Load Agent Adjusted Limit*
S=*Structural Weight Limit*
T=*Takeoff Weight Limit, based on climb or runway limited weights*
E=*Enroute Considerations*
D=***Dispatcher Adjusted Limit***

Snow, Ice, or Slush Corrections

Flap Settings authorized: 11 or 17° ***only***
Separate slush/wet snow tables for ½" and ¼" of slush
No correction required for 1½" or less of dry snow
Use ½" Slush table for:
- Ice Conditions
- 1½ to 3" of dry snow

Contaminated Runways

A runway should be considered contaminated when:
More than 25 percent of the required field length, within the width being used, is covered by:
- Standing water, slush, or wet snow deeper than 1/8" (3 mm)
- Dry snow deeper than 1 inch (25 mm)
- Ice

If a runway is contaminated:
- Takeoff is not authorized with a tailwind
- Takeoff is not authorized with
  - More than 1/2 inch of wet snow, slush, ***OR***
  - Standing water, ***OR***
  - More than 4 inches of dry snow
- Maximum thrust must be used
  - If ART (MEL item 73-8) is inoperative use Reserve thrust. Standard thrust is not authorized
- Both thrust reversers must be operative
- APU will be used for takeoff, if operative.
  - APU Air Switch is **OFF**
  - APU Bus Switches are **ON**)
- Takeoff not authorized with chunks of hardened snow or ice
- Corrections to V1 and maximum weight allowances are made by dispatchers and sent to aircraft

Performance, TAKEOFF 40.1
Repeated on page 22 of this Study Guide

Automatic Relay Summary

Note: *The table on this page is designed to assist in sorting out the various automatic functions and which actions are required to arm, disarm, and activate each. No attempt is made to include the actions resulting from activation, as these are generally self-explanatory. The Auto Pack Shutdown, for instance shuts down the packs when it is armed and activated. For each column, the page number for expanded notes in this Study Guide is given at the top of the column.*

Where multiple conditions are listed all, must be satisfied unless indicated otherwise	Auto Pack Shutdown	Auto Spoiler	Auto Brakes	ART	ATR	Anti-Skid	AutoSlats
Operating Manual:	AIR 20.1	FLT-C 20.5	LAND 20.5	ENG 20.8	ENG 20.6	LAND 20.3	FLT-C 20.9
TAKEOFF Study Guide:	Page 24	Page 38	Page 48	Page 34	Page 34	Page 47	Page 39
Engines—one or both running Pack Supply Switches (one or both) in **HP BLD OFF** or **AUTO** AUTO selected—Air Conditioning Shutoff Switch Differential Pressure (cabin to ambient) below 1.3 psi	ARM						
Differential 13th stage pressure (delta P) of 70 psi or more as failed engine spools down	ACTIVATE						
Airborne through approx. 3000 feet AFL	DISARM						
Air Conditioner Shutoff Switch to **OVRD**	RESET						
Squeeze & Raise Spoiler lever		ARM					
Reverse thrust selected		ACTIVATE					
Flaps positioned to less than 26° Auto Brake **ARM-DISARM** Switch to **ARM** Anti-Skid **ARM-DISARM** Switch to **ARM** Spoilers Stowed T. O. Position on **AUTO BRAKE** selector			ARM				
Spoiler handle out of full forward (auto or manual) 1 Second delay **MAX**, 3 Second delay **MIN** or **MED**			ACTIVATE 1 - 3 sec delay				
RTO Above 70 Knots			ACTIVATE MAX				
RTO Below 70 Knots			ACTIVATE MIN				
Throttles retarded below 22° Brake pedals NOT depressed Spoilers deployed (automatically or manually)			ACTIVATE				
Flaps raised to less than 26° with speed above 70 knots ***OR*** *(Below 70 knots, Raising flaps to less than 26° does not discontinue ABS)* Throttle—Either advanced beyond 22°* ***OR*** ARM-**DISARM** Switch to **DISARM** ***OR*** Brakes—Either brake pedal depressed more than 25% ***OR*** Stowing Ground Spoilers* *Note: *Both items inhibit autobrakes—they will be reapplied if criteria reversed.*			DE-ACTIVATE (Released)				
Self test complete and **READY** Light ON Both engines above 64%				ARM			
30% N_1 Difference				ACTIVATE			
Before Slat Retraction—Both engines retarded below 58% N_1				TEMP DISARM			
Power on both engines advances above 64%				RE-ARM			
Slat Retraction				DISARM			
Takeoff—Flight director in take-off (TAK OFF) mode Altitude—Radio Altitude over 350 feet Power—Both engines operating below Go-Around power Note: *If engine failure occurs below 350 feet, ATR does not engage when altitude reaches 350 feet. (See added notes page 28)*					ARM		
Vertical speed decreases below 0 fpm for five seconds OR EPR drop ≥0.25 *and* N_1 loss ≥7% on the same engine					ACTIVATE		
Any pitch mode selected on DFGC					DISARM		
Anti-Skid Switch to **ARM** Main Wheel Spin Up						ARM	
Skid Sensed by Anti-Skid Computer						ACTIVATE	
Landing Gear not Down Arming Switch OFF Parking Brakes Set Slow Speed Taxi (Approx. 10 knots)						DISARM	
Speed below 240 knots Flap/Slat Handle in 0 to 13 T/O (EXT) range							ARM
Either stall-warning computer detects approach to stall							ACTIVATE

LANDING: *(Only Differences are listed below; Arm/Disarm/Activation conditions above apply where appropriate.)*

	Auto Pack Shutdown	Auto Spoiler	Auto Brakes	ART	ATR	Anti-Skid	AutoSlats
Spoiler lever raised		ARM					
Flight Spoilers: Main Wheel spin up ***OR*** Nose Strut Compression Ground Spoilers Added: Weight on Main Gear and Throttles Idle		ACTIVATE					
Throttles forward for go-around after spoiler deployment		DISARM	DISARM				
Anti-Skid Armed ABS Switch in a LDG range (Not **OFF**) ABS Arming switch to **ARM**			ARM			ACTIVATE/ DISARM *as above*	
Spoilers activated—see above: **MAX**—1 second delay after spoilers activated **MED** or **MIN**—3 sec delay after spoilers activated			ACTIVATE				

AutoBrakes

Must be armed (If operative) for any of the following:
Runway
Less than 7,000'—Auto Brake
Contaminated—standing water/snow/slush/ice
Weather
RVR less than 4000 or visibility less than 3/4 mile
CAT II or III Approaches—Autobrakes armed if operable
Braking conditions reported less than good
Minimum Stopping Distance situation—use of MAX req'd
Recommended: landing with gusty winds or crosswinds
OM Vol. 1, APP, LDG, G/A 10.3, 40.2, 40.5

Flaps 40 Landing

- **Normal flap setting for landing** is flaps 28
  Conserves fuel vs. Flaps 40
  Flaps 28 normally used:
  —On dry runways more than 7,000' long
  —At airport elevations over 6,000' MSL
  —May be required at other airports
- **Autoland** authorized for flaps 28or 40.
- **Flaps 40 required** at airport elevations 6,000' MSL or less when any of the following apply:
  —Tailwind
  —Wet / slippery runways
  —Braking action less than good
  —Captain's judgment
- **Flaps 40 helps reduce** overall stopping distance

OM Vol. 1 APP-LDG_G/A 10.4-10.5 , 50.2-50.3

Maximum Flap and Gear Speeds

0-13°—280	26-27°—200	Gear Extension 300 / .70M
14-20°—240	*MD-83—205*	Extended 300 / .70M
21-25°—220	28-40°—195	Retraction 250 / .70M
	MD-83—200	OM Vol. 1, LIM 10.6

Unofficial Guide ** Training Only

— 5 —

Fold

Standardized Actions

Callouts—Bold Italics Actions-Non Bold PM Actions GreenGear

Takeoff	2 Engine Go-Around	Takeoff Engine Failure	1 Engine Go-Around
Autothrottle—ON *Thrust Set* *80 Knots—Check* *V₁* *Rotate* *V₂* *V₂+10* *Positive Rate,* *Gear Up*	*Go-Around* *Set Go-Around Thrust* *Flaps – 15* (or *Flaps 11* as required) *Positive Rate,* *Gear Up* *Set and Arm Missed Approach Altitude*	*Autothrottle—ON"* *Thrust Set* *80 Knots—Check* *V₁* *My Aircraft* *Rotate* *V₂* *Positive Rate,* *Gear Up* *Runway Heading, HDG SEL*	*Go-Around* *Set Go-Around Thrust* *Flaps 11* **VREF + 5 Minimum** *Positive Rate,* *Gear Up* *Runway HDG, HDG SEL* *Set and Arm Missed Approach Altitude*
Passing 400' AGL Minimum			
HDG SEL or *NAV*	*HDG HOLD, NAV* or *HDG SEL* *Set Speed 200*		
Passing 500' AGL Minimum			
Autopilot ON (if desired)			*Set Speed 250*
> 1000 AFL		**EOAA**	**> 1000 AFL**
Half Rate *Climb Power* *Flaps Up* (on schedule)	*Half Rate* *Climb Power* *Flaps Up* (on schedule) *Speed Select*	*Altitude Hold* *If Engine Fire or Severe Damage:* Memory Items Autothrottle...OFF Throttle Affected Engine..Confirm ....IDLE *Flaps Up* *Slats Retract* (On Schedule)	*Altitude Hold* *Flaps Up* (On Schedule)
0/EXT *Slats Retract*			**0/EXT** *Slats Retract*
		0/EXT+ 20	
Clean Min Maneuver IAS, Bank 30 **2500' AFL** *IAS 250* or *SPD SEL 250* or *PERF* or *VNAV*		*IAS* *A/C Override* *SET MCT* *Checklist*	*IAS* *A/C Override* *SET MCT* *Checklist*

OM Vol. 1, General 15.4

Unofficial Guide ** Training Only

— 6 —

Fold

MD-80 Briefings & Notes

Flight Attendant Briefing

Number of F/A's on board
Known delays
Ground—Short-taxi & safety demo considerations
Turbulence
Weather and turbulence forecast for route
Unexpected turbulence
Notification
Required action in the event of a significant encounter
Service suspension at F/A's discretion
When able, call CA to report cabin situation / injuries
If no access to interphone, remain seated, avoid injury
Resuming Duties
Once cleared to resume call flight deck to report any injuries or abnormalities.
Security items
E6 Cabin Items
Life vest demo--overwater segment
Gen Declarations/Customs Immigration forms (if req'd).
Cockpit Access
Crew meals
Cockpit Door
Cabin Door Handles—Push Down
Any other unusual issues relevant to the flight
FM Part 1, page 7.1-3

TEST Briefing

T—TYPE of emergency
E—EVACUATION will be accomplished after landing
S—Evacuation SIGNAL to be used
T—TIME to landing

Unofficial Guide ** Training Only

— 1 —

Cut ✂

Operational Notes and Lists

Disclaimer: *On this page and the one following page are lists and consolidated information which have operational relevance. These lists are intended as a starting point for consolidating each pilot's personal notes, and the reproducible pages here are*

For Training Purposes Only

This material is provided in an effort to help in consolidating policy guidelines for planning purposes.

It is not an attempt to replace Operating Manual guidance from which these notes are derived.

Takeoff Briefing

Pilot Flying
Rejected T/O
- <80—Unusual Indications
- >80—Eng **Failure**, **Fire**, **PWS**, **Unsafe**

Contingencies
- Wet Runway — See PERF TAKEOFF 10.2
- Departure
- Apt Specific Eng Out
- TO Alternate
  - <330NM, must accommodate engine out approach
- T/O Weather Considerations
- R/W Surface Considerations
- Terrain
- Other Taxi/T/O Variables & NOTAMS

FM Part 1, page 8.2-1; OM Vol. 1, STARTING 10.9

Standard Thrust Not Permitted

Maximum Thrust Required
- Tailwind
- Runway contaminated with standing water, slush, snow, ice
- Windshear reported / expected
- TOW > ATOW (*Can get new TPS*)
- Improved Performance used
- When FM-II Airport Advisory requires Max
- MEL / CDL items
  - With weight corrections ***not*** included in TPS
  - With prohibitions of Standard Thrust
- Engine Anti-ice used; TPS not planned for A/I

Reserve Thrust Required
- ART System inoperative, and Performance Manual does not authorize standard thrust

Note: *It is the Captain's responsibility to ensure that no condition exists that prohibits use of standard Thrust even though it appears on the TPS.*

Performance-TPS 10.4
OM Vol. 1, Taxi-Takeoff 30.6, LIM 10.18

Unofficial Guide ** Training Only

— 2 —

Fold

Single Engine Taxi Guidance

Not authorized when:
- Jet blast will be excessive
- Ramps & taxiways slippery
- Icing conditions
  - —No significant precipitation (snow, sleet or freezing rain) occurring that could adhere to / collect in engine inlet.
  - —The engine inlet will be checked by Maintenance or authorized deicing personnel prior to restart.

Before Takeoff APU recommended for SE taxi
After Landing APU not required, but consider:
- All galleys and overwing heaters shed
- AC Load limitation is 1.0—may need to turn off non-essential equipment
- If cabin temperature uncomfortable—restart APU

Cautions and Considerations
- Avoid excessive jet blast in ramp areas with high weight
- Size of aircraft following may be a factor
- Monitor L and R fuel tank quantities for imbalance
- Restart may be accomplished while taxiing at Captain's discretion — OM Vol. 1, TAXI-TAKEOFF 10.2; AFTER LANDING-Parking 10.3

Common Sense Issues—May not be advisable when:
- Braking Action Less than good
- MEL items involving normal / back-up modes of
  - ◆Generators ◆Hydraulics ◆Steering & Brakes
- Proximity of the gate to departure / arrival runway
- Tight turns in confined areas expected

Approach Setup—WARM-V

Initial Setup—"WARM"
- <u>W</u>eather requirements & NOTAMS—Check
  - Include crosswinds, visibility, published minimums
- <u>A</u>pproach—Plan and brief approach
- <u>R</u>NP—Check chart & Set As Required for the Approach
- <u>M</u>CP—When cleared for the approach—Select APP Mode
  - Set FAF altitude when on intercept or a portion of the approach

Note: *These are techniques extracted from various sources*

Unofficial Guide ** Training Only

— 3 —

Fold

Passing FAF Actions

Passing FAF Actions—MAC-V
- <u>M</u>issed Approach Altitude—Set when 300' below MAP
- <u>A</u>PPR Light—Green (RNAV Approaches)
- <u>C</u>learance—Tower Call Passing FAF
- <u>V/S</u>—Set as required

Note: *These are techniques extracted from various sources*

Final Approach Deviation Callouts

Airspeed .... +10 / -5 knots*
Rate of Descent .... <2000'—More than 2000 fpm
.... <1000'—More than 1000 fpm*
.... Inside FAF—>1000' fpm

Inside Final Approach Fix
- Localizer Deviation —1/3 dot deviation.......... "COURSE"
- Glide Slope — ½ dot deviation.......... "GLIDESLOPE"
- Non-ILS / RNAV (GPS / GNSS) deviation
  - — 1 Dot lateral course deviation.......... "COURSE"
  - — ½ Scale on VDI (100' AFL or less).......... "GLIDESLOPE"
- VOR deviation exceeds 2º.......... "COURSE"
- NDB deviation exceeds 5 º.......... "COURSE"
- LOC deviation exceeds ⅓ dot on PFD.......... "COURSE"

PM Directs "GO AROUND" if PF not correcting
APPROACH, LANDING, GO-AROUND 20.4-20.5

Unofficial Guide ** Training Only

— 4 —

Cut ✂

Takeoff Briefing
Single Engine Taxi
Standard Thrust Not Permitted
Approach Setup
Passing FAF Actions
Autobrakes
Flaps 40 Landing Recommended

CKLST | Table of Contents | Overhead Panel | Fwd Panel Ctr Console

Titles Currently Available

By Rick Townsend

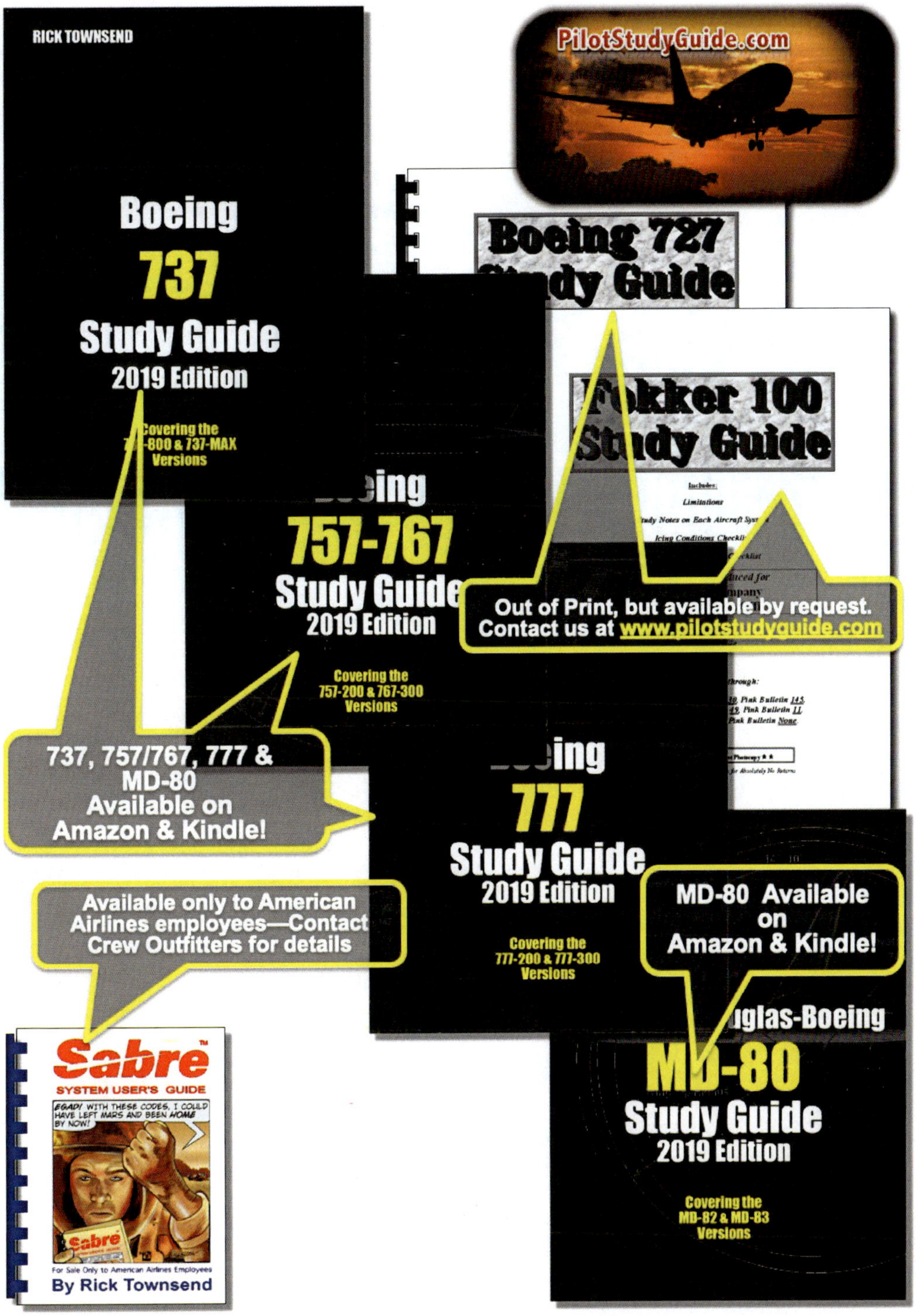

To purchase any of these titles, go to Crew Outfitters locations at the Flight Academy Gift Shop, or at DFW Airport Terminal C Operations
To purchase online, go to www.crewoutfitters.com
To order by phone, call **800-874-1204**

This book is available through Amazon in both Kindle and in print format!

Giving Back

A portion of the proceeds from every book help support some worthy charities. Here are two of the charities your book purchase supports:

Kenya Kids Can!

Feeding and teaching world changers of tomorrow

Kenya Kids Can is a non-profit working in the Rift Valley of Kenya. They feed students and construct solar powered computer centers to improve education and bring hope to needy students. Please browse the website to learn about what can be done to truly change lives.

www.kenyakidscan.org

Snowball Express

GARY SINISE FOUNDATION

SNOWBALL EXPRESS

PRESENTING SPONSOR American Airlines

Serving the *children* of our fallen military heroes

Snowball Express Could Use Your Help!

Snowball Express honors America's fallen military service members who have made the ultimate sacrifice since 9/11 by

- Humbly serving the families they left behind
- Championing their children's future success by creating opportunities for joy, friendship, and communal healing
- Connecting these families to one another.

Since 2006, the mission of Snowball Express has been a simple, yet profoundly important one: Providing hope and new happy memories to the children of military fallen heroes who have died while on active duty since 9/11. In December each year we bring children together from all over the world for a four-day experience filled with fun activities, like sporting events, dances, amusement parks and more.

Nationally, Snowball Express provides comprehensive support programs for fallen families that are focused on transition and connections to community resources, healing and wellness, peer engagement, education and personal/professional development programs.

www.snowballexpress.org

Made in United States
North Haven, CT
04 November 2026

21734477R00035

IRIS
MAGAZINE

TABLE OF CONTENTS

A VISIONARY ENTREPRENEUR AND ADVOCATE FOR CHANGE

Dr. Iris Wright stands at the forefront of entrepreneurial innovation and community advocacy, with a dynamic portfolio that includes Author Iris Wright & Publishing, Wright Circle of Care, formally known as Caring Hearts Telecare, Wright Way Logistics, Wrights Holdings, and her influential platforms: Author Iris Wright, Real Talk with Iris, and Black Diamond Chronicle Magazine. Married for 15 years, Dr. Wright is the matriarch of a blended family with six children and three grandchildren. Eight years ago, she relocated from Delaware to Newport News, VA, seeking new opportunities and fresh beginnings.

Iris's entrepreneurial spirit and passion for service emerged early in life, laying the foundation for her wide-ranging career. After a decade in restaurant management and various roles, her true calling in healthcare became clear at age 28, leading her to launch a mission centered on compassionate care. In 2019, inspired by Steve Harvey's Jump, she left traditional employment to build her own legacy—launching Wrights Virtual Services. Her journey expanded with the creation of Caring Hearts Senior Living in 2021, later rebranded as Caring Hearts Telecare in 2022, earning consecutive nominations for Best Home Care of the Year in 2023 and 2024.

Dr. Wright's impact extends

beyond business. In July 2022, she was invited to speak at a Leading Age seminar on Telehealth in Home Care, showcasing her progressive approach to elder care. Her leadership has earned her the 2025 Phenomenal Woman Award, the Top 100 Innovator & Entrepreneur Award, 3x Presidential Lifetime Achievement Awards, Forbes' Top 20 Entrepreneurs to Watch in 2024, the CREA Global Award, the Empowered Diva Award, the I Inspire Global Award, and the America Award. She is also a Christlet Award Winner for Black Diamond Chronicles and an International Impact Award Winner for her Injustice series, Black Diamond Chronicles, and Being Me Is Super.

Additional recognition includes the Community Leader Award from ACHI Magazine, Community Votes Awards in both 2023 and 2024, the 2025 Icon Award, the I Inspire Global Award, and two Honorary Doctorates in Healthcare Administration and Business. Dr. Wright won the 2025 Entrepreneur Award.

As an author, Dr. Wright uses her voice to uplift others. Her bestsellers include Taking Your Power Back, From Brokenness to Brilliance, Black Diamond Chronicles, Injustice Vol I, Injustice Vol II, Injustice Vol III, and her children's books Being Me Is Enough and Being Me Is Super, which encourage self-love and confidence in young readers. Her upcoming solo works, Injustice: My Story and Evolving the Storm, explore her personal journey of triumph and transformation. With new projects like The Impact Leader Featuring Kevin Harrington (The Original Shark from Shark Tank), and Teen Mom to Entrepreneur launching in 2025, her influence through writing continues to grow.

Beyond business and authorship, Dr. Wright is a fierce advocate for affordable home care and fair reimbursement for seniors. Her Injustice Movement, launched in 2023, provides legal support to the wrongly accused and works to reunite families torn apart by injustice. She is currently involved in a clemency campaign for a man who has served 27 years of an 80-year sentence for a non-violent crime.

Dr. Iris Wright's deep-rooted commitment to justice, innovation, and community care makes her a driving force for change. Through her ventures, voice, and vision, she continues to inspire others to walk boldly in purpose and create lasting impact.

Connect with Dr. Iris Wright:

- ***Website:*** iris-wright.com
- ***Facebook:*** Author Iris Wright
- ***Instagram:*** @author_iris_

1. What impact are you most proud of creating through your work, and why does it matter?

I am most proud of empowering people to realize possibilities for their lives that they once thought unattainable.

In business, healthcare advocacy, and community leadership, I help people move forward through guidance, encouragement, and information. The impact may involve advising an entrepreneur, advocating for healthcare awareness, or sparking conversations about equality in communities.

What matters most to me is when someone says, *"Because of that conversation, I moved forward,"* or *"I believed in myself a little more."*

These moments prove that real influence is about enabling others to grow. My purpose is to create spaces where people feel empowered to pursue ideas, advocate for themselves, and transform their futures.

That impact matters because empowerment is generational. When one person gains knowledge, confidence, or access, it doesn't stop with them. It influences families, communities, and future leaders.

If my work helps people realize their potential, build something meaningful, or advocate for themselves and others, then it serves its purpose.

2. Who or what has been most affected by your work?

The people most affected by my work have been individuals in the communities I serve, especially aspiring entrepreneurs, women seeking leadership opportunities, and families navigating access to information in business and healthcare.

Through mentorship, advocacy, and community engagement, I've had the opportunity to support people who simply needed guidance, encouragement, or direction. Many individuals have great ideas or goals but lack the resources or confidence to move forward. Helping someone clarify their vision, develop a strategy, or recognize their own potential has been one of the most rewarding parts of my work.

Women have been deeply impacted by the conversations and leadership platforms I've been involved in. When women see other women stepping into leadership, embracing entrepreneurship, and advocating for change, it creates a ripple effect. It encourages them to believe that they can do the same.

Community members have benefited from my work in healthcare awareness and social justice. Addressing inequities, fostering dialogue, and improving access strengthen communities. Informed and empowered people make impactful decisions for future generations.

Ultimately, the impact of my work is measured by the people who feel more confident, more informed, and more empowered to pursue opportunities because someone believed in them or spoke up on their behalf. That ripple effect is what makes the work meaningful.

3. How do you define meaningful impact in your industry?

For me, meaningful impact is measured by transformation, not just recognition.

Impact in my fields is meaningful when people gain opportunities, knowledge, and resources that improve their lives. It is about transforming conversations and initiatives into real progress.

Meaningful impact is an entrepreneur building a sustainable business, a community learning about healthcare access, or women embracing leadership. Seeing these changes means people feel supported and empowered.

I also believe meaningful impact is sustainable. It goes beyond a moment or a headline. It creates systems, partnerships, and conversations that continue to benefit people long after the initial work is done.

In my experience, impact is often quiet. It's the person who implements the advice you shared. The community member who feels heard and supported. The young woman decides to pursue leadership after seeing someone who looks like her succeed in it.

When people leave a space stronger, more informed, and more empowered than when they arrive, that's meaningful impact to me. It means the work is creating real change, not just visibility.

4. What results best reflect your progress or success?

The results that best reflect my progress are the influence I have on others and the opportunities created through my work.

For me, success isn't just about titles or recognition, though receiving an Honorary Doctorate was meaningful. What truly reflects progress is seeing people grow because of the guidance and advocacy I provided.

Seeing entrepreneurs gain clarity and confidence to build sustainable businesses—and watching those businesses support families or communities—powerfully reflects impact.

Results include increased awareness of healthcare access and social equity. When people are more informed, and leaders act on equity, real movement occurs.

Mentorship is another sign of success. When women step into leadership, speak confidently, or pursue new opportunities, it shows that encouragement matters.

Progress is measured by growth—personal and collective. When people feel empowered to act and lead because of my work, that is true success.

5. Can you share a specific obstacle and the results that followed?

One of the biggest obstacles I faced was deciding to leave the security of my job and fully step into entrepreneurship. At the time, it was a major risk. Stability can make you feel safe, but I knew there was a bigger purpose calling me to do more.

The decision became clearer after I read Jump by Steve Harvey. That book challenged me to stop waiting for perfect timing and start trusting my preparation and faith. Still, making that leap was not easy. There were moments of uncertainty, financial concerns, and pressure to build something from the ground up. Yet, that obstacle became a turning point.

Stepping into entrepreneurship has let me broaden my influence beyond a traditional role. I focused on business, healthcare advocacy, and community leadership, which opened doors to mentorship, collaboration, and growth initiatives.

That leap strengthened my confidence and reinforced the value of purpose over comfort, leading to my Honorary Doctorate in recognition of my contributions to business, healthcare, and the community.

Looking back, that obstacle taught me that the moments that feel most uncertain are often the ones that drive the greatest growth. Sometimes the biggest breakthrough begins with the courage to take one step forward.

6. How do you stay focused on results without losing sight of your mission?

My focus comes from always remembering why I started.

Results matter. In my work, progress must be measurable. But I ensure that results never outweigh the mission.

My mission centers are on service, empowerment, and expanding access. When making decisions or setting goals, I ask: Does this reflect my purpose? If yes, I know I'm on the right path.

I also stay grounded by staying connected to the people the work is meant to serve. When you're actively listening to communities, mentoring entrepreneurs, or advocating for better healthcare awareness, it keeps the work personal. It reminds you that behind every goal or initiative are real people whose lives can be affected.

I maintain balance by focusing on long-term impact rather than quick wins. Sustainable change requires patience and consistency. I measure success by enduring results.

Staying focused on results yet true to the mission requires discipline and clarity. A clear purpose helps filter opportunities and stay aligned with what matters.

7. What does innovation mean to you beyond buzzwords?

Innovation means solving real problems in ways that measurably improve people's lives.

"Innovation" often means trendy changes, but real innovation solves problems and creates better outcomes. True innovation finds smarter, more effective solutions to old challenges.

In the industries I work in, business, healthcare advocacy, and community leadership, innovation often means bridging gaps that people have learned to accept as normal. It means asking questions like: Why is this system not working for everyone? What resources are missing? How can we make access easier and more equitable?

Sometimes innovation is not about inventing something completely new. Sometimes it's about improving how things are done. It might create clearer pathways for entrepreneurs to start sustainable businesses or help communities better understand healthcare resources, enabling them to make informed decisions.

For me, innovation also requires listening. When you listen to the experiences of people in your community, you begin to see where change is needed. That insight allows you to develop practical, sustainable solutions rather than just theoretical ones.

Most importantly, innovation should lead to impact. If an idea sounds impressive but doesn't improve outcomes for people, then it's just a buzzword.

True innovation creates access, opportunity, and progress that people can feel in their everyday lives.

8. How do you encourage innovation within your organization or work?

I encourage innovation by creating an environment where ideas are welcomed, questions are encouraged, and people feel safe sharing new perspectives.

One of the most important ways innovations grow is through open dialogue. When people know their voices are valued, they're more likely to offer ideas that can improve processes, programs, or services. I try to create spaces where team members and collaborators can speak honestly about what is working, what is not working, and what could be done differently.

I also encourage innovation by staying connected to the communities we serve. Real innovation often comes from listening to the needs and experiences of people on the ground. When we pay attention to those insights, we're able to develop solutions that are practical and relevant rather than theoretical.

Another important part of encouraging innovation is being willing to adapt. Not every idea will work the first time perfectly, and that's okay. I try to foster a culture where learning and improvement are part of the process. When people know that thoughtful experimentation is supported, they're more willing to think creatively.

I also believe in collaboration. Some of the best ideas come from bringing together different perspectives across business, healthcare, and community leadership. When people with different experiences work together, they often see solutions that one person alone might miss.

Ultimately, encouraging innovation means creating a culture that values curiosity, problem-solving, and continuous growth. When people feel empowered to think beyond the usual way of doing things, meaningful progress becomes possible.

9. How are emerging technologies, including AI, influencing your approach?

Emerging technologies, including artificial intelligence, are expanding how we access information, make decisions, and deliver services. For me, these tools are not replacements for human leadership, but powerful resources that can strengthen how we work and serve communities.

In business and community initiatives, technology helps streamline processes, analyze information more quickly, and improve communication. AI allows us to gather insights that can guide better strategies and more informed decision-making. Instead of spending excessive time on routine tasks, leaders can focus more on planning, relationship-building, and impact.

Technology is also helping expand access to knowledge. Entrepreneurs, community leaders, and professionals now have tools that allow them to research, learn, and develop ideas faster than ever before. This creates new opportunities for innovation and growth, especially for people who may not have had traditional access to mentorship or training.

At the same time, I believe it's important to approach technology responsibly. Tools like AI should enhance human judgment, not replace it. In areas such as healthcare advocacy and community leadership, empathy, ethics, and lived experience remain essential. Technology can provide data and efficiency, but people provide wisdom and context.

My approach is to embrace technology to amplify impact. When used thoughtfully, emerging technologies can help leaders reach more people, improve systems, and create solutions that are more responsive to communities' needs.

The goal is balance, using innovation to move forward while keeping human purpose at the center of the work.

10. What does a strong strategy look like when it's executed well?

A strong strategy, when executed well, is clear, focused, and produces measurable results.

It begins with a well-defined purpose. Everyone involved understands the goal, why it matters, and how their role contributes to achieving it. When people have clarity around the mission, decisions become more aligned, and progress becomes easier to track.

Execution also requires consistency. A great strategy is not just an idea on paper. It's a plan that is followed through with discipline, accountability, and adaptability. There are milestones, timelines, and clear responsibilities so that progress can be measured and adjusted when necessary.

Another sign of a strong strategy is flexibility. Even the best plans must adapt to changing circumstances. Leaders who execute well pay attention to feedback, evaluate results, and refine their approach without losing sight of the overall mission.

Communication is also essential. Teams need regular updates, transparency, and collaboration to stay aligned. When communication is strong, it builds trust and ensures everyone is moving in the same direction.

Most importantly, a well-executed strategy creates real impact. It leads to meaningful outcomes, whether that's stronger businesses, better access to resources, or more informed communities. The success of a strategy is ultimately measured by the difference it makes in the lives of the people it's meant to serve.

DR. MONIQUE L. RODGERS

Dr. Monique L. Rodgers is an internationally award-winning author, ordained prophet and pastor, motivational speaker, and transformational leader whose life and work are grounded in faith, purpose, and prophetic insight. She is widely recognized for her dynamic voice in Christian thought, literary creativity, and commitment to equipping individuals to discover and live out their God-given destinies. She is a literary genius, book birther and The GOAT of Books.

Dr. Rodgers holds advanced degrees in Global Leadership and ministry training from institutions such as Colorado Technical University and Oral Roberts University, reflecting a strong foundation in both academic study and spiritual formation. Her academic and ministerial journey fuels her passion for guiding others through personal transformation, spiritual depth, and purposeful living.

As the founder and CEO of WriteOn, LLC and Shooting Stars Publishing House, Dr. Rodgers champions authors and creatives by helping them bring meaningful, impactful content into the world. Through her coaching, publishing guidance, and keynote presentations, she blends faith-based wisdom with practical strategy, inspiring writers and leaders to refine their craft and amplify their voice. Her contributions extend to collaborative works and bestsellers, including titles like Called to Intercede, I am Black History, Called to Lead and Accelerate.

Dr. Rodgers has been featured on national and international platforms as a speaker and literary authority. Her work has earned significant recognition, including the Presidential Lifetime Achievement Award (2023), Iconic Author of the Year 2024, Glim Awards Iconic Education and Youth Award 2026, She has also received honors in peer-reviewed lists of influential authors, and invitations to minister and speak across diverse audiences. She has traveled extensively, engaging with communities in Africa, Mexico, Peru, Jamaica, Haiti, and Thailand, reinforcing her commitment to global impact and intercultural ministry.

Known for her candid storytelling, uplifting encouragement, and spiritual acuity, Dr. Monique L. Rodgers centers her work on restoring hope, igniting purpose, and equipping others to walk confidently in their calling. Whether through publishing, coaching, ministry, or speaking engagements, her mission remains steadfast: to transform lives through the written and spoken word and to raise up generations who live boldly for God's Glory

1. What impact are you most proud of creating through your work, and why does it matter?

I am most proud of helping people reclaim their voice and see purpose beyond their pain. Through books, coaching, and community initiatives, I've helped aspiring authors, entrepreneurs, and faith-driven leaders turn their stories into platforms for healing and empowerment. It matters because when one person heals or succeeds, entire communities are strengthened.

2. Who or what has been most affected by your work?

Emerging authors, faith-based leaders, and entrepreneurs seeking clarity and confidence have been most impacted. Many have gone from uncertainty to publishing books, launching businesses, and stepping into leadership roles.

3. How do you define meaningful impact in your industry?

Meaningful impact is not just sales or recognition—it's transformation. When someone's life trajectory changes because they discovered their purpose or gained confidence through my work, that's real impact.

4. What results best reflect your progress or success?

Publishing hundreds of books, mentoring authors, creating profitable platforms, and seeing my clients build sustainable brands are all signs of success. But the greatest result is seeing others thrive.

5. Can you share a specific obstacle and the results that followed?

Overcoming homelessness and financial hardship was a major obstacle. That journey taught resilience and empathy, which now fuel my

mission to help others build stability and purpose through entrepreneurship and authorship.

6. How do you stay focused on results without losing sight of your mission?

I regularly revisit my purpose service, empowerment, and faith-centered leadership. Results matter, but the mission guides every decision.

7. What does innovation mean to you beyond buzzwords?

Innovation means creating solutions that genuinely solve problems and open doors for others, not just following trends.

8. How do you encourage innovation within your organization or work?

I create spaces where new ideas are welcomed, experimentation is encouraged, and learning from failure is embraced.

9. How are emerging technologies, including AI, influencing your approach?

AI allows writers and entrepreneurs to work faster and smarter. I use it ethically to accelerate content creation, marketing, and education while keeping authenticity at the center.

10. What does a strong strategy look like when it's executed well?

A strong strategy aligns vision, systems, and people so progress happens consistently without chaos.

11. How do you balance long-term vision with short-term strategy?

I break long-term goals into actionable milestones while ensuring each step supports the larger mission.

12. What strategic lesson has shaped your leadership the most?

Collaboration accelerates growth. Building partnerships and empowering others produces greater results than working alone.

13. How has your leadership style evolved?

I've shifted from doing everything myself to empowering teams and leaders to step into their strengths.

14. What mindset shift has been critical to your growth?

Understanding that setbacks are setups for growth, not signs to quit.

15. What advice would you give leaders who want to create lasting impact?

Lead with purpose, invest in people, stay adaptable, and build something that outlives you by empowering others to succeed.

GINA MARIE CARACCI

Host of Mindset, Health, Empowerment:
The Unfiltered Trainer
Author of The Badass Blueprint on Amazon
CEO of GinaMarieFit,LLC
Founder of Sweating Together For A Cause & IMPACT Society

Gina Caracci is a health and wellness entrepreneur, coach, and speaker dedicated to helping women build strength from the inside out. She is the host of The Unfiltered Trainer, a no BS mindset and fitness podcast with over 100,000 downloads, where she shares real talk, sustainable wellness strategies, and tools for personal growth.

Gina is also the author of The Badass Blueprint, a practical guide to building confidence, discipline, and lasting transformation without burnout. As CEO of GinaMarieFit, LLC, she brings more than 11 years of personal training and group fitness experience, helping thousands of clients transform their health and mindset. Through her brand, she also operates a user generated content and lifestyle wellness media business, partnering with national companies to create authentic, conversion driven content that builds brand awareness and drives sales.

She is the founder of Sweating Together For A Cause and IMPACT Society, where she blends fitness, business, and philanthropy to unite communities, support local causes, and create meaningful, lasting impact.

Please answer the following 15 IRIS Magazine interview questions, centered around Impact, Results, Innovation, and Strategy:

1. What impact are you most proud of creating through your work, and why does it matter?

The impact I am most proud of creating is building a space where people feel safe to show up exactly as they are without shame, without comparison, and without feeling like they have to earn their worth.

From the beginning, I knew I did not want to build a brand that only measured success in followers or finances. I wanted to create something deeper.

Through my content, my podcast, my fitness classes, and our community events, my goal has always been to make people feel seen, supported, and empowered.

Through our charitable wellness events, we have raised thousands of dollars for local nonprofits while giving them visibility they might not otherwise receive. That matters because it proves fitness can be about more than aesthetics. It can be about service, connection, and real change.

Inside my classes and online community, I have watched people transform not just physically, but mentally and emotionally. They gained confidence, strength, and belonging. No one felt like they had to compete or look a certain way to deserve a spot in the room. They were welcomed exactly as they were.

That natural culture of inclusion is what I am most proud of. Because when people feel safe, they grow. When they feel supported, they stay. And when they feel accepted, they finally believe they are capable of more.That is the kind of impact that lasts far beyond any metric.

2. Who or what has been most affected by your work?

The people most

affected by my work are my clients and my listeners.

My clients, both online and in person, experience more than just physical results. Yes, they get stronger, healthier, and more confident in their bodies, but what truly moves me is the mind, body, and soul transformation that happens alongside it. I have watched women set boundaries, leave toxic environments, pursue new careers, rebuild their self worth, and completely change how they show up in their lives.

That ripple effect is everything.

And in a beautiful way, their growth impacts me just as much. Their trust challenges me to lead better, show up stronger, and continuously evolve. They inspire me to be a leader they feel proud to follow and empowered by. It is not just coaching. It is a shared transformation.

My listeners are just as deeply affected and honestly, they are the ones driving the dream forward.

Through the podcast, I get messages from women who say an episode helped them make a hard decision, set a boundary, start their fitness journey, or finally believe in themselves again. Knowing that my voice can reach someone on a walk, in their car, or during a tough season of life and make them feel less alone is something I never take lightly.

They are the reason I work so hard to create thoughtful, relevant, and impactful content every week.
At the end of the day, it is not about numbers. It is about people feeling seen, supported, and capable of more than they thought possible. If my work helps even one person change the trajectory of their life, then it matters.

3. How do you define meaningful impact in your industry?

I define meaningful impact as creating change that lasts beyond the moment.

In the wellness and fitness industry, it is easy to focus only on aesthetics, numbers, or surface level transformations. But for me, meaningful impact has never been about sets, reps, or before and after photos. It is about how someone's life shifts because of the experience.

That belief sits at the forefront of my entire business and honestly dictates every decision I make.
My wellness events are never just workouts. They always carry a deeper purpose. Whether we are raising awareness for mental health, domestic violence, housing and food insecurity, supporting local charities, or bringing women together to feel less alone, we use fitness as the vehicle for something bigger. Movement becomes connection. Sweat becomes service.

Community becomes change.

Through my podcast, I take the same approach. We talk about the conversations many people avoid. Burnout. Confidence. healing. boundaries. identity. Mental health. I lead with honesty and authenticity so listeners feel seen, supported, and valued no matter what they have experienced. If someone finishes an episode feeling stronger mentally or emotionally, that is impact to me.

At the core, my work has never been a "me" thing. It is a *"we"* thing. Meaningful impact is building spaces where people feel safe, empowered, and capable of more and then watching that confidence ripple into every part of their life. Because when wellness improves, everything improves. That is the kind of change that actually lasts.

4. What results best reflect your progress or success?

For me, success has never been measured by revenue alone. It has always been measured by reach, connection, and real life change. The results that best reflect my progress are the ones that show impact at scale.

My podcast statistics are a tangible example of that growth. What started as a simple idea and a microphone has grown into a global movement with over 75,000 downloads, listeners in 78 countries, and hundreds of cities around the world. That tells me the conversations we are having matter. It means the messages are resonating far beyond my immediate circle and reaching people who need support, clarity, or encouragement in their everyday lives.

But just as meaningful to me are the thousands of clients and event attendees I have had the privilege to work with in person and online. Every class taught, every event hosted, every client coached represents a human being who trusted me to be part of their journey. Watching those mind, body, and soul transformations and seeing the ripple effects into their confidence, careers, relationships, and overall well being is what I truly consider success.

Because at the end of the day, progress is not just about numbers. It is about how many lives you can positively touch. The more I can give back, the more successful I feel.

5. Can you share a specific obstacle and the results that followed?

Two years ago, I faced an obstacle that completely changed the trajectory of both my life and my business. I was in a car accident where I was hit on the driver's side and sustained a traumatic brain injury. Ironically, it happened during one of the biggest growth seasons of my career. In the year leading up to the accident, I had been recognized with a 40 Women to Watch award, earned a Forbes Entrepreneur of Impact nomination, and was nominated as a community favorite for the Tampa Urban Excellence Awards. Everything felt like it was accelerating. And then, overnight, everything stopped.

I went from building momentum to struggling with basic cognitive, physical, and emotional tasks. I had to pause my goals, my events, and much of my work to focus entirely on recovery. Over the next year and a half, I went through multiple forms of therapy to regain my abilities while also learning to adapt to permanent changes. It was humbling. It was frustrating. And at times, it was incredibly isolating. But it also reshaped me in the best way possible.

That season forced me to slow down, reevaluate what truly mattered, and redefine success. I developed a deeper resilience, a stronger work ethic, and a renewed sense of purpose. When I returned, I came back more intentional, more grounded, and more committed to impact than ever before.

I stepped back into in person events, expanded into new networks, and poured fully into my podcast and personal brand with a different perspective. Not just to grow, but to serve. What resulted was not just a comeback. It was alignment. Now, everything I build comes from a place of gratitude, clarity, and purpose, not pressure. And that has made both my work and my impact stronger than it ever was before.

6. How do you stay focused on results without losing sight of your mission?

For me, my mission has always been impact first and everything else second. I stay focused by building and diversifying offerings that genuinely serve people, whether that is through my events, coaching, or my podcast, making sure each one provides real support, connection, and transformation instead of just chasing trends. The numbers and revenue follow naturally, but I measure success by the lives changed, not the metrics on a dashboard.

7. What does innovation mean to you beyond buzzwords?

Innovation, to me, is about changing narratives and challenging the status quo so we can create real, lasting impact in our communities, not just following trends or chasing buzzwords.

It means being bold enough to stand up for what is right, have honest conversations, and build spaces that are more inclusive, supportive, and purpose driven. It also requires the humility to pivot, evolve, and adapt as the world changes so the work we do stays relevant and truly serves the people we are here to help.

8. How do you encourage innovation within your organization or work?

I encourage innovation by staying closely connected to the real needs of the people we serve and letting that guide everything we create. Within Sweating Together For A Cause and IMPACT Society, we host events that go beyond fitness by addressing mental health, community challenges, and charitable causes so our work continues to evolve with

purpose, not routine. Through my podcast and my online brand, I lean into honest, sometimes uncomfortable conversations and create content around hard pressing issues that others may avoid. For me, innovation is less about doing something flashy and more about having the courage to adapt, listen, and lead in ways that truly make a difference.

9. How are emerging technologies, including AI, influencing your approach?

AI influences my work in an intentional and supportive way, not as a replacement for creativity or connection. I primarily use it behind the scenes for editing, video development, and streamlining production so I can save time and operate more efficiently.

That extra time allows me to focus on what really matters, which is building relationships, creating meaningful content, and showing up fully for my community through my events and the podcast. For me, technology should enhance the human experience, not replace it, so I use AI as a tool to amplify my impact, not automate my voice.

10. What does a strong strategy look like when it's executed well?

A strong strategy, when executed well, feels grounded in peace, clarity, and alignment rather than chaos or constant pressure. When you have a clear plan and back it with intentional action, gratitude, and purpose, the stress fades and your energy becomes focused instead of scattered. You are no longer reacting. You are moving with direction and confidence. That alignment not only drives stronger results and greater impact, it creates a sense of fulfillment and genuine happiness along the way.

KIMBERLY SCHMITZ

Kimberly Schmitz is a Fractional CMO and Marketing Strategist, Founder of TINK Marketing & Design and the Brand Luminary League, based in the Denver Metro Area, Colorado. With over twenty years of experience as a marketing communications professional, she has crafted strategies for businesses ranging from early-stage startups to corporate enterprises.

She specializes in helping small and mid-sized founders build a clear marketing foundation from day one, using the same proven methodology she has applied with large organizations, made affordable and simple to implement. She is also the founder of Brand Luminary League, a monthly membership program helping solopreneurs and early-stage entrepreneurs develop their story and make marketing work for them.

When she is not thinking about marketing, you can find her practicing or teaching Pilates, and spending time with her family and two dogs. Learn more at www.tinkmarketing.com.

KIMBERLY SCHMITZ

1. What impact are you most proud of creating through your work, and why does it matter?

Honestly, the thing I am most proud of is the moment a founder finally sees their own story as an asset. So many of the people I work with come to me feeling overwhelmed by marketing, like it's this big, complicated thing they are not qualified to do. But marketing is not brain surgery. It just needs to be made simpler so founders can stop spinning on it and focus on everything else their business needs from them. When we do the foundational work together, and they realize, "Oh, this is actually about who I am and why I built this," something shifts. They gain clarity. They talk about their business differently, with conviction and intention, because they finally know exactly what they are saying and why. That transformation matters to me because it is not just a marketing win. It shifts them from Operator Mode to Owner Mode.

2. Who or what has been most affected by your work?

Solopreneurs and small business founders, without question. These are people who are full of passion and vision but often paralyzed by the marketing piece. They think they need a big budget or a big team to get it right, and a lot of them have been burned by jumping straight into tactics before they had a strategy. But there is also something bigger that drives me, and that is the health of our local business community. When small businesses grow, communities grow. A founder gets clear on their story, starts showing up with confidence, and suddenly they are not just building a business, they are becoming a real part of the fabric of their community. What I have seen is that when a founder gets clear on their story and their foundation, everything else gets easier. Their messaging clicks. Their confidence grows. Their business starts to move. And the ripple effect goes well beyond them. That is what keeps me going.

3. How do you define meaningful impact in your industry?

For me, meaningful impact is when marketing stops feeling like a chore and starts feeling like an authentic expression of who someone is and what they are building. The marketing industry, especially digital marketing, tends to make small business owners feel like they are always behind, not doing enough, or doing it wrong. I want to be the person who demystifies it, who says: you already have everything you need, let us just build the right framework around it. When a founder walks away from our work together with a clear plan they actually believe in, that is meaningful impact.

RESULTS

4. What results best reflect your progress or success?

The results I am most proud of are not always the flashy ones. Yes, I have helped businesses grow revenue and build brand recognition. But what really reflects success to me is when a founder who started our work together without a clear message ends up with a brand story they can tell confidently in any room. Or when a business owner who was throwing money at random tactics finally has a strategic plan that makes sense for where they are right now. Clarity and confidence are the results I chase, because they are the ones that make everything else possible.

5. Can you share a specific obstacle and the results that followed?

One that comes to mind is working with a founder who had been in business for a few years but felt completely stuck. They had tried different marketing approaches, spent money on things that did not work, and were honestly starting to question whether their business was viable. When we dug into the foundational work, the real issue was not their tactics. It was that they had never clearly defined who they were for or why their approach was different. Once we built that foundation, their messaging became specific and compelling, they started attracting the right clients, and the noise in their head about marketing quieted down. That is the kind of turnaround that reminds me why this work matters.

6. How do you stay focused on results without losing sight of your mission?

I come back to the why regularly. And honestly, my own founder story is what keeps me most grounded. I know what it feels like to be in the weeds of a business, trying to figure out marketing without a clear strategy or a roadmap. That lived experience is exactly why I do this work, and reconnecting with it reminds me who I am here to serve. My mission has always been to make marketing accessible and less intimidating for founders who are building something real. When I feel myself getting pulled toward the shiny new tactic or the latest trend, I ask whether it actually serves the people I am here to help. Results and mission are not at odds with each other in my work. When I stay true to the mission, the results follow. The moment I start chasing results for their own sake, I lose what makes my approach distinctive.

INNOVATION

7. What does innovation mean to you beyond buzzwords?

To me, innovation is about finding a better way to create a real outcome for real people. It does not have to be technology-driven or groundbreaking. Sometimes the most innovative thing I do is take a marketing methodology that has worked for large organizations and translate it into something a solo founder can actually use without a team or a six-figure budget. That translation, making a powerful strategy accessible, is where I find the most creative energy. And I think there is something genuinely innovative happening right now in how we think about human storytelling in a world that is increasingly automated. Everyone has access to the same AI tools. Everyone can produce polished content at scale. What they cannot produce is your story, your lived experience, your specific why.

The founders who recognize that their human story is the one thing no algorithm can replicate are the ones who will stand out. To me, that is one of the most exciting and important marketing innovations of this moment. Innovation is not about being first. It is about being genuinely useful in a way that others have not figured out yet.

8. How do you encourage innovation within your organization or work?

I encourage innovation by giving myself

permission to just get something out there. It does not have to be perfect. It has to be a start. I think one of the biggest blockers for founders, and honestly for me too sometimes, is waiting until something feels completely ready before putting it into the world.

But you cannot get feedback on something no one has seen yet. With Brand Luminary League, I am constantly trying new ways to teach founder story development, putting ideas out, watching what lands, listening to where people get stuck, and then improving from there. That feedback loop is where the real innovation happens. I pay close attention to where people struggle, because that is usually where the opportunity for a better approach lives. The best frameworks I have built did not come from a perfect plan. They came from a founder saying "I just do not get it yet" and me going back to figure out a clearer, simpler path. Give yourself permission to try. Then let what you learn shape what comes next.

9. How are emerging technologies, including AI, influencing your approach?

I use AI tools and I think they are genuinely useful for the right things. Where they shine is in execution: drafting content, scaling production, saving time on the tactical work. But I am very intentional about where the human work has to happen first. No AI tool can uncover a founder's story or articulate the specific belief that makes their brand different. That is human work, and it has to come before the technology. So my approach is: build the foundation first, get crystal clear on who you are and why you are in business, and then use AI as an amplifier. Without that foundation, you are just automating noise.

STRATEGY

10. What does a strong strategy look like when it is executed well?

A strong strategy, executed well, feels almost invisible to the founder because everything makes sense together. There is no random tactic, no disconnected campaign, no "we should probably do more on social media" without a clear reason.

When strategy is working, every marketing decision connects back to a core story, a clear goal, and a clear vision. The founder knows who they are talking to, what they are saying, and why it matters to that specific person. And marketing starts to feel less like a burden and more like a natural extension of the business itself. That is the state I am always working toward with clients: marketing that is clear, purposeful, and actually theirs.

Kimberly Schmitz is a Fractional CMO/ Marketing Strategist and owner of TINK Marketing & Design LLC, based in the Denver Metro Area, Colorado. She works with small to mid-sized founders to build the marketing foundation that creates clarity, connection, and momentum from day one. She is also the founder of Brand Luminary League, a monthly marketing membership program helping early-stage entrepreneurs develop their story and make marketing easier for themselves. You can learn more at www.tinkmarketing.com

HUMAN-CENTERED MARKETING IN A TECH-DRIVEN WORLD

Why your founder story is still the most powerful differentiator you have, and why no algorithm can replicate it.

By Kimberly Schmitz

Right now, somewhere, a founder is watching an AI tool generate a month's worth of social media posts in forty-five seconds. The captions are polished. The hashtags are optimized. The posting schedule is set. And yet something essential is missing. No prompt can produce the one thing that matters most: the reason why that business exists at all.

We are living through a genuine inflection point in marketing. Automation tools, AI-generated content, and data-driven personalization have lowered the barriers to execution so dramatically that virtually any business can look and sound like a sophisticated brand. Which means that looking and sounding polished is no longer a differentiator. In a landscape flooded with algorithm-perfect content, what actually cuts through is human truth.

The Noise Problem

For founders just getting started, this shift is both a challenge and a tremendous opportunity. The challenge is real: you are building credibility from scratch, and the temptation is to use AI tools to look more established than you are. The opportunity is far more powerful. The one asset no competitor can replicate is you: your lived experience, your values, the specific moment that made you decide to build something.

Consumers have grown remarkably good at distinguishing between content that was produced and content that was felt. They may not be able to articulate it, but they feel it. There is a subtle flatness to a message that was optimized rather than authored. Trust, which is the actual currency of small-business marketing, is not built through a consistent posting cadence. It is built through consistency of character.

Your Story Is Strategy

In my work with founders who are in the early stages of building their businesses, one pattern shows up again and again: they underestimate the strategic value of their own story. When you are just getting started, it is easy to think you need to look established before you can earn trust. So, founders rush toward tactics, logos, and campaigns before they have ever answered the most important question: why does this business exist, and why should anyone care? But the story is not the preamble to the strategy. At the foundation stage, the founder's story is the strategy.

Here is why that matters: when a prospective customer compares your business to a competitor, they are often evaluating very similar offers at comparable price points. Features can be copied. Services can be bundled.

Guarantees can be matched. What cannot be copied is the specific why behind your business: the problem you lived before you solved it, the belief that shapes every decision you make, the customer you are uniquely equipped to serve because you have been where they are.

That is not sentiment. That is positioning. And it is positioning that no competitor, human or machine, can manufacture.

Using AI Without Losing Yourself

None of this is an argument against AI tools. Used

well, they are genuinely useful for repurposing content, scaling production, testing messaging variations, and freeing up the hours founders spend on execution so they can invest more in strategy.

The mistake is using AI as the source rather than the amplifier.

Think of it this way: AI is extraordinarily good at transmitting a message efficiently. It is not capable of originating a message that carries emotional weight and earned credibility. The work of clarifying who you are, why it matters, and who you are for has to come from you first. Once that foundation is built, automation becomes a powerful delivery mechanism. Without it, you are simply automating noise.

Where to Start

If you are a founder who is just beginning to build your brand and have been focused on getting something out into the world before clarifying the story underneath it, two foundational exercises are worth working through before your next move.

Your Purpose and Mission: The Why Behind the Business

Your mission is not a tagline. It is the answer to why you showed up to do this work at all. To uncover it, ask yourself:

- Where do you excel? What comes naturally to you that others find difficult or overwhelming?
- What problem did you experience personally before building this business? The answer to this question is often the most powerful thing you can say to a prospective customer, because it signals that you have lived what they are living.

Your Vision: The Future You Are Building Toward

Your vision is the destination. It gives your brand direction and gives your audience something to believe in alongside you. To find it, ask:

- What are you genuinely passionate about? Not what sounds good, but what you would work on even if no one was watching.
- Where can you make a real difference? Who changes because of the work you do, and how?

These are not branding exercises. They are the foundation that makes all your marketing, AI-assisted or otherwise, actually work.

The Enduring Advantage

Marketing technology will keep evolving. The tools available today will seem primitive in five years. But the fundamental human dynamic that drives purchasing decisions, the trust built through authentic connection, is not going to change. People buy from people. They buy from businesses whose values they recognize. They buy from founders whose stories resonate with their own.

The founders who will build something lasting in this next chapter of marketing are not the ones who adopt every new tool fastest. They are the ones who start with clarity about who they are, build their brand on that foundation, and then use every available tool to say it better, louder, and to exactly the right audience.

In a world where anyone can publish anything instantly, your human story is the rarest thing you have. It is also the most powerful marketing asset you will ever own.

Kimberly Schmitz is a Fractional CMO/Marketing Strategist and owner of TINK Marketing & Design LLC, based in the Denver Metro Area, Colorado. She works with small to mid-sized founders to build the marketing foundation that creates clarity, connection, and momentum from day one. She is also the founder of Brand Luminary League, a monthly marketing membership program helping early-stage entrepreneurs develop their story and make marketing easier for themselves. You can learn more at www.tinkmarketing.com

REV. DR. JOANNE ANGEL BARRYCOLON

Rev. Joanne Angel BarryColon is the visionary Founder of **Wholistic Fitness NY LLC.** With over 40 years of experience in health, fitness, nutrition, and wellness, she has spent the last 15 years integrating various healing modalities into her unique practice. Rev. Joanne's holistic approach empowers individuals to achieve wellness in mind, body, and spirit through personalized guidance and transformative techniques.

An expert in multiple disciplines, Rev. Joanne is a **wholistic integrative teacher, certified wholistic personal trainer, Reiki Master, medical astrologer,** and **master numerologist.** She holds a **B.S. in Health, Fitness, and Nutrition from Empire State University,** further enhancing her credentials in the wellness space.

Rev. Joanne extends her knowledge and healing philosophy to a global audience through her show, **Joanne's Healing Within,** which airs live on **Strong Island Television** on the first Monday of each month. The program is also available on platforms like **Roku, iHeartRadio, Audible,** and **Spotify,** where she provides insightful content on holistic health, self-empowerment, and personal transformation.

A prolific author, Rev. Joanne has penned six books, coauthored more than a dozen others, and contributed articles to publications such as **Your Success Magazine** and **The Authors Porch Magazine.** She currently writes for **Brainz Magazine** and is the creator of the **Chakra Balance Numerology Cosmic Energy Forecast Deck.** Her work blends ancient wisdom and modern wellness practices, offering her audience practical tools for spiritual growth and empowerment.

Rev. Joanne's achievements have been recognized by **P.O.W.E.R. (Professional Organization of Women of Excellence Recognized),** where she was honored as a **Woman of Empowerment.** Additionally, she was celebrated as **Woman of the Month** in November. As the founder of the **Women Empowering Women Private Group,** she creates a safe and nurturing space for women to connect and grow. The group meets on Zoom on the second Monday of each month, fostering community and support among its members.

A core part of Rev. Joanne's mission is to help women over 40 build strength and discipline through a revolutionized approach to fitness and wellness. Her methods emphasize more than just physical training; they empower women to cultivate confidence, courage, and resilience as they align with their soul's purpose. Over her career, she has transformed the lives of more than **20,000 women,** guiding them toward healthier, stronger, and more empowered futures.

For more information, you can reach Rev. Joanne at Healingwithin76@gmail.com or visit her websites:

www.wholisticfitnessny.com
www.youtube.com/@WholisticFitnessNY

1. What impact are you most proud of creating through your work, and why does it matter?

The impact I'm most proud of is holding sacred space for women over 40 to release emotional weight without dieting while gaining the confidence and courage to walk away from their 9-to-5, outdated relationships, and old habits and beliefs. This allows them to fully embrace their soul purpose.

It matters because every woman's purpose was designed to support the next generation of leaders and innovators. When one woman shines her light, she gives permission for the next woman to see hers. Women were never meant to dim their light or stand in darkness.

2. Who or what has been most affected by your work?

Women over 40 who are ready to evolve. I teach them the tools to put down the baggage and release emotional weight so they can stand boldly in their greatness.

With Mars in the tenth house in Aries in my natal chart, my mission is clear — to transform women and build leaders.

3. How do you define meaningful impact in your industry?

Meaningful impact on health, fitness, nutrition, and wellness goes far beyond numbers on a scale. It is strength, longevity, mental health, and full-spectrum wellness — physical, emotional, mental, and spiritual transformation.

It fulfills the quality of L.I.F.E. Living in Freedom Everyday where a woman is independent physically, mentally, and emotionally, free from preventable health issues.

4. What results best reflect your progress or success?

The greatest reflection of my success is witnessing clients evolve into authors, speakers, writers, and entrepreneurs transforming not just their bodies, but their emotions, mindset, spirit, and soul.

5. Can you share a specific obstacle and the results that followed?

For 22 years, I operated a storefront business without fully recognizing my value and worth as a business owner. The Pandemic became a turning point. I transitioned into a fully virtual model.

What felt like an obstacle became an expansion. My clientele increased. I began traveling to speak. I launched a magazine available in digital and print formats. Today, I work four hours a day while generating a full-time income. That shift required courage and clarity.

6. How do you stay focused on results without losing sight of your mission?

My daily health and wellness practice anchors me. When I embody what I teach, alignment naturally follows.

7. What does innovation mean to you beyond buzzwords?

Innovation is authentic in motion. It is thinking beyond conventional structures and creating from imagination and divine inspiration.

As an Aquarius, I am naturally a visionary, collaborator, thinker, and researcher. Innovation allows me to support clients in aligning with their soul purpose in both strategic and intuitive ways.

8. How do you encourage innovation within your work?

I shift the focus from weight loss to strength. I guide clients to fuel their bodies with high-quality protein, healthy carbohydrates, fiber, healthy fats, and proper hydration with electrolytes.

We quiet the *"monkey noise"* and prioritize mental health.

My mantra Release Emotional Weight Without Dieting represents the whole person. The *"W"* stands for body, mind, emotions, spirit, and soul.

My approach is non-conventional. It is revolutionary.

9. How are emerging technologies, including AI, influencing your approach?

At first, I resisted AI. Now I embrace it fully. I use it to enhance content creation, graphic design, and strategic visibility, expanding my message and growing my business.

10. What does a strong strategy look like when executed well?

A strong strategy produces transformation that leads to evolution and once you evolve, you do not return to who you were.

11. How do you balance long-term vision with short-term strategy?

As an Aquarius, I naturally see beyond what is directly in front of me. The short-term strategy is simply the stepping stone that supports and aligns with the long-term vision.

12. What strategic lesson has shaped your leadership the most?

Fully embracing my Mars in Aries in the tenth house placement transformed my leadership. Mars is the driving force of the chart. The tenth house represents soul purpose. Aries ask: Who am I? Why am I here? What legacy will I leave?

With this placement, leadership is not optional. I am here to initiate, execute, and stand firmly in my authority.

Astrology tells a story composed of 12 zodiac signs, 12 houses, and 10 planets. The planets are the verbs what's happening. The signs are the adjectives how it happens. The houses are the nouns where it happens.

Astrology is a language and it is a navigation system for your life.

13. How has your leadership style evolved?

The deeper I understand my Mars placement, the more decisive and courageous I become. I make bold decisions without hesitation. I take steps that may feel uncomfortable and move forward anyway.

Most importantly, I lead with intuition over ego.

14. What mindset shift has been critical to your growth?

Understanding that I am not meant to lead everyone, and that not everyone is meant to be in my circle.

I confidently enter the rooms I am invited to and gracefully exit those that are not aligned.

15. What advice would you give leaders who want to create lasting impact?

Follow your Soul's Purpose. Learn your Astrological Make-up your built-in navigation system for this lifetime.

Would you drive to an unfamiliar destination without using GPS? Then why travel through life without activating your Astrological Make-Up Navigation System (A.M.N.S.)?

TREVA BENDY

Treva Bendy is the Founder and CEO of The Training Lounge, LLC, a learning and development company centered on career mobility through structured skill development. With more than 20 years of experience in instructional design, corporate training, and organizational performance, she has designed enterprise learning solutions, supported large-scale training initiatives, and strengthened workforce capability for Fortune 500 companies across the energy and technology industries.

Building on that foundation, she now leads two distinct pathways that help professionals move forward with clarity and execution.

The Instructional Design Mastery Accelerator is a 12-week program for educators and professionals preparing to transition into instructional design. Participants learn how to analyze business needs, design structured learning solutions, collaborate with stakeholders, build storyboards, and develop interactive eLearning experiences. They graduate with a completed eLearning portfolio that demonstrates real instructional design skills and positions them to compete across corporate, healthcare, government, and technology environments.

Course Mastery for Entrepreneurs is a guided program for coaches and subject matter experts who

want to transform their expertise into a structured digital course. Participants define their target audience, organize their knowledge into a clear curriculum, create engaging lessons, and prepare for launch using measurable standards that support sustainable revenue.

Treva is also the author of Embracing Change: Overcoming Obstacles of Career Transitions, a published book that addresses the emotional and practical realities of career change. It speaks directly to the fear, identity shifts, and financial concerns that often surface during career transitions. Through interactive exercises and structured guidance, readers are led through a clear process to evaluate their current position, define what they want next, and build a plan to move forward with intention.

Through her programs and published work, Treva helps individuals turn experience into demonstrated capability that leads to measurable results.

1. What impact are you most proud of creating through your work, and why does it matter?

The impact I am most proud of is creating structured pathways for career mobility, so people leave with marketable skills, tangible proof of capability, and a clear strategic direction.

Many professionals face layoffs or career stagnation and lack a clear strategy for repositioning themselves. That is why I built programs that convert experience into marketable capability, whether that means transitioning into instructional design or building a revenue-generating digital course.

What makes this meaningful is that it replaces uncertainty with structured execution.

Instead of reacting to job disruption, my clients build skills that position them competitively. This matters because workforce shifts are accelerating, and long-term stability now depends on clarity, discipline, and strategic action.

2. Who or what has been most affected by your work?

The individuals most affected by my work are mid-career professionals, educators, and subject matter experts who know they are capable but lack a clear strategy for what comes next. Some come with experience but without a defined process to align their skills with current market expectations. Others are stepping into new areas and need to build technical capability from the ground up.

What changes for them is not just knowledge. It is competence that can be demonstrated. They learn how to build work that meets defined standards rather than relying solely on past experience. Some develop the technical skills required to create effective eLearning. Others learn how to structure their expertise into a clear, organized, and outcome-driven course. In each case, the result is a measurable capability they can stand behind.

The impact extends beyond individuals. When graduates enter corporate, healthcare, government, or technology environments, they are prepared to perform at a higher level, benefiting organizations as well. The ripple effect influences teams, departments, and overall performance.

3. How do you define meaningful impact in your industry?

Meaningful impact in learning and development is not measured by content volume or completion rates. It is measured by a demonstrable change. Did someone secure a role because they could show proof of capability? Did an expert launch a structured digital program that generates sustainable income? Did a training initiative improve performance within an organization?

Impact must be tied to execution. Strategy without measurable results is theory. In my industry, innovation is not about tools. It is about designing structured learning solutions that close the gap between learning and performance.

For me, meaningful impact is when education changes someone's market position. When skills translate into advancement, income, or improved performance, that is impact.

4. What results best reflect your progress or success?

The results that reflect my progress are not limited to job placements or course launches. They are visible in the quality of work my participants produce. When someone who once felt stuck can clearly articulate their value and demonstrate it through completed work, that reflects success.

Educators leave with eLearning portfolios that meet corporate expectations and position them for new roles. Entrepreneurs leave with a structured digital course built around their expertise and designed to deliver measurable value. When that work leads to advancement, revenue, or stronger professional positioning, that reflects sustained success.

5. Can you share a specific obstacle and the results that followed?

One of the most defining obstacles in my journey was being laid off while working in instructional design. I had already transitioned from Industrial Engineering into the field and was building momentum. The layoff forced me to pause and reevaluate. I wasn't questioning my capability, but I did have to reconsider how I positioned myself and what kind of impact I wanted to make next.

Instead of treating it as a setback, I treated it as a strategic reset. I assessed where my skills created the most value and leaned into both instructional design and change management. That decision led me to a Change Management Manager role, where I oversaw instructional design and broader change initiatives. It expanded my perspective from designing learning solutions to leading transformation efforts.

That experience sharpened my thinking about career mobility. I saw how quickly circumstances can shift, even for high performers. More importantly, I learned the value of structure during uncertain moments. The frameworks I used to reposition myself became the foundation for The Training Lounge. What started as a disruption became refinement.

6. How do you stay focused on results without losing sight of your mission?

I stay focused by remembering why people enroll in the first place. They're not looking for more information. They're looking for change. Many come in feeling stuck, undervalued, or unsure how to move forward. That starting point matters, but it cannot be where they stay.

My mission is career mobility. That means helping people build real, marketable skills and produce work that proves what they can do. So, I hold my programs to a clear standard. Are they leaving with tangible work, such as a completed

eLearning portfolio or a structured digital course? Can they demonstrate new capability? Has their professional position improved?

If those outcomes aren't visible, I reassess the structure and make the necessary adjustments. Results are how I measure whether the mission is being fulfilled. When someone can show proof of growth, not just talk about it, I know the program is working.

7. What does innovation mean to you beyond buzzwords?

Innovation, to me, is not about adopting new tools. It's about producing better outcomes. In learning and development, innovation happens when skill development is directly tied to measurable results. That means reducing wasted effort, closing the gap between learning and performance, and aligning education with real workforce demand.

If a program cannot deliver tangible capability or improve someone's professional position, it is not innovative. Real innovation strengthens execution and produces results that can be demonstrated, not just discussed.

8. How do you encourage innovation within your organization or work?

I encourage innovation by starting with outcomes, not content. Before anything is delivered, I define what must change and what must be produced. What will participants build? What capability should exist at the end that did not exist at the beginning?

I regularly assess whether the structure of my programs reflects current workforce expectations. If industries evolve, the standards must evolve as well. Innovation is not about constant reinvention. It is about disciplined refinement. Within my work, that means setting clear performance standards, measuring output, and improving the program where gaps appear.

9. How are emerging technologies, including AI, influencing your approach?

Emerging technologies, including AI, are accelerating the creation of content. But speed alone does not create value. I see AI as a productivity tool, not a strategic substitute. It can support research, help structure initial course outlines, and streamline workflow, but it cannot replace instructional judgment, stakeholder alignment, or structured design.

In my programs, I teach participants how to use technology intentionally. The focus is not on letting AI create the work, but on using it to strengthen clarity, improve quality, and increase efficiency. When used correctly, AI enhances execution and allows more attention to be placed on design decisions and measurable outcomes.

10. What does a strong strategy look like when it's executed well?

A strong strategy begins with knowing exactly what must be built. There should be no confusion about the outcome, the expected quality level, or the next step. In my work, that means defining success before anything is developed. If the goal is a job-ready portfolio, then the assignment must result in a completed eLearning portfolio that demonstrates real instructional design skill. If the goal is launching a digital course, then each milestone must move someone closer to a defined audience, a structured curriculum, and a deliverable ready to launch.

When it's executed well, every action serves a purpose, and there is tangible work to show for it.

11. How do you balance long-term vision with short-term strategy?

My long-term vision is centered on career mobility through structured learning. But vision without execution is only intention. So, I break that vision into distinct phases. What needs to be built this quarter? What capability must be strengthened this month? What outcome should be achieved in the next two weeks?

Each short-term action has to support the larger direction. If it doesn't move the mission forward, I remove it. The long-term vision gives the work purpose. The short-term plan keeps it practical and measurable. Balancing both means staying focused on what matters now without losing sight of where we're headed.

12. What strategic lesson has shaped your leadership the most?

The most important lesson I've learned is that clarity is more powerful than speed. Early in my career, I saw what happens when direction isn't clearly defined. People stay busy, but meaningful progress stalls.

I learned to slow down long enough to define the objective clearly and set measurable standards before moving forward.

When the outcome is defined, execution becomes faster and more focused. That lesson shapes how I lead today. I prioritize clear expectations, defined deliverables, and accountability tied to results. Strategy without clarity creates noise. Clarity creates momentum.

13. How has your leadership style evolved?

My leadership style has evolved to be more intentional about listening so I can lead with greater precision. Not everyone enters my programs at the same stage in their career. Some are navigating layoffs. Some are shifting industries. Others are building something new for the first time. Early on, I focused primarily on structure and outcomes. Over time, I realized that structure is most effective when it supports people where they are.

That realization shaped how I lead inside my programs. We provide step-by-step guidance, structured mentorship, and weekly Q&A sessions so participants can ask questions and receive support when needed. While each person's starting point may differ, the expectations remain consistent. Everyone is held to the same measurable outcomes.

The structure is clear. The support is intentional. That balance allows individuals to grow from their current level while still producing tangible work that meets the standard.

14. What mindset shift has been critical to your growth?

A critical mindset shift for me was realizing that experience alone isn't enough. What worked five years ago may not meet today's expectations. That forced me to stop relying on past success and start regularly evaluating whether my skills and programs still align with current demand.

Now, I don't assume relevance. I assess it. I consider industry expectations, hiring standards, and market shifts, and adjust as needed. Staying relevant isn't about reacting to trends. It's about having the discipline to make adjustments before the market forces you to.

15. What advice would you give leaders who want to create lasting impact?

Start by understanding the people you are serving. Know what they actually need and what it will take for them to reach the outcome. Assumptions weaken the impact. Clarity strengthens it.

Define the result clearly and make it measurable. If you cannot identify what success looks like, you cannot protect it. Lasting impact requires structure, accountability, and the discipline to evaluate whether the work is producing real change.

And when it isn't, be willing to adjust the structure without lowering the standard. Impact is sustained when leaders stay true to the mission and are honest about the results.

Link to The **Training Lounge website:** www.TheTrainingLounge.com
Link to purchase my book, **Embracing Change: Overcoming Obstacles of Career Transitions:** https://www.3gpublishinginc.com/shop-lo16Z/p/embracingchange.

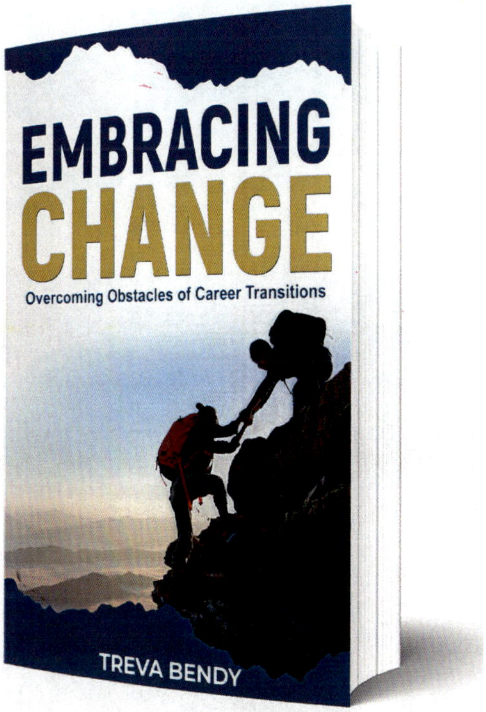

REGGIE BENDY JR

REGGIE BENDY, JR AND BOOK DETAILS

Reggie Bendy, Jr. is a 17-year-old high school senior and the author of The Tales of Christmas Hallows. He is an emerging young writer who took something he genuinely loves and turned it into a story he could share with the world. Reggie represents a new generation of storytellers who understand that magic is not just something you wait for. It is something you create.

Christmas has always held a special place in his heart. Not just the traditions or the celebration, but the joy that comes with them. He has always loved how Christmas brings families together and creates moments people remember for years.

Inspired by watching his mother write and publish

her own books, he did not simply talk about writing one day; he did it. He committed to writing his own book. That determination reflects who he is.

Reggie wrote The Tales of Christmas Hallows because he wanted readers, especially young readers, to experience the joy of Christmas year-round. He wanted them to believe that the magic of Christmas should never fade and is not limited to one month on the calendar. Through his story, he hopes readers see that the spirit of Christmas lives in kindness, courage, and the way we show up for one another every day.

He approaches storytelling the same way he approaches life: with heart, focus, and intention.

Outside of writing, Reggie enjoys music, loves biology, spends time fishing with his father, and exploring new technology. He's naturally curious and drawn to innovation, always learning and discovering tools that expand his creativity.

At his core, Reggie believes stories should do more than entertain. They should connect families, spark conversations, and remind us of what truly matters. His goal is simple: to write stories that make readers smile, reflect, and feel inspired long after the final page.

About the Book

The Tales of Christmas Hallows

The Tales of Christmas Hallows is a heartfelt holiday adventure that invites readers into a town where the spirit of Christmas is much more than a holiday; it's a way of life.

When young Damien and his family move to the mysterious town of Christmas Hallows, what seems like a fresh start quickly turns into a journey of discovery filled with wonder, unexpected friendships, bravery, and self-belief.

As he navigates new friendships and faces unexpected challenges, Damien begins to realize that Christmas Hallows holds more than traditions. It holds secrets that test his courage and change the way he sees the magic around him. The story appeals to readers of all ages, weaving fantasy and emotional depth into a narrative that feels both timeless and uplifting. It's a book families can read together and return to year after year.

Inside every copy, Reggie includes a unique QR code that unlocks a heartwarming surprise. This personal touch deepens the connection between the story and the reader and reflects Reggie's love for creativity and innovation.

The Tales of Christmas Hallows is more than a seasonal book. It's a reminder.
A reminder that joy is intentional.

That courage can be quiet but powerful.
Those traditions endure when we protect them.

And that the light of Christmas should never fade.

The Tales of Christmas Hallows is available in both paperback and hardback formats, offering readers a magical escape perfect for gift-giving and yearlong reading traditions.

Link to purchase The Tales of Christmas Hallows: https://www.3gpublishinginc.com/shop-lo16Z/p/thetalesofchristmashallows-4tp43

Link to Reggie Bendy, Jr. website: www.StoryHallows.com

What Readers Are Saying

Readers on Reggie's site describe The Tales of Christmas Hallows as:

- A hidden gem full of heart and joy
- A story that becomes a treasured family tradition
- A magical adventure that both children and adults can enjoy
- An inspiring reminder that Christmas magic can live in our hearts all year long

ELIZABETH MINER

ELIZABETH MINER | MINER ENTERPRISE, LLC

Elizabeth Miner is the founder of Miner Enterprise, LLC, an executive coaching and leadership advisory firm specializing in emotional intelligence for executives and founders.

She is the author of The Entrepreneur Advantage: Emotional Intelligence for Building and Growing Your Business, a speaker, and a recognized voice in leadership development and entrepreneurship.

She brings a distinctive lens to her work, drawing on her background as a former Apple employee, paralegal, and multi-venture founder. She is a community builder who has led numerous business groups worldwide.

Elizabeth's overarching methodology centers on the conviction that developing emotional intelligence is the highest-value investment a leader can make.

1. What impact are you most proud of creating through your work?

When people strengthen their emotional intelligence, the effect does not stop with them. The leader I work with becomes more confident, more influential, and a sharper communicator. That improvement ripples outward to every person they lead, every room they walk into, every relationship they carry.

At a time when misunderstanding, polarization, and uncertainty are the backdrop of nearly every conversation, every boardroom, and every negotiation, empathy is not a soft skill; it is a requirement. Effective leadership begins with knowing your audience deeply, and everything else, how you communicate, how you innovate, how you build, follows from that foundation.

The impact I am most proud of is that the work does not end in the room where we do it. One leader changed is a team changed; sometimes, a whole organization changes. That reach is what gets me up in the morning.

1. Who has benefited most from what you do, and how have you seen that impact in action?

The obvious answer is the leaders themselves. They leave my work more effectively, more confident, and better able to navigate the complexity their roles demand. But the more interesting answer is everyone around the leader.

We have all seen what poor leadership costs. Talented people leave. Culture erodes. The bottom line follows. What gets less attention is the reverse: when a leader grows, the whole environment shifts. People feel connected. They understand their role. They commit to the mission in a different way. People will rise or fall to the level of leadership.

The moment I know the work truly matters is when I see it change an organization's culture, not just one person's behavior. That ripple effect is the proof of concept. It is also the reason the work is worth doing, where it is resisted most.

2. What results best demonstrate the success of your work so far?

The clearest results show up in how leaders operate under pressure. When an executive who once defaulted to command-and-control starts leading with curiosity, asking better questions, and creating space for their team to own outcomes, that is a measurable shift. Retention improves. Decisions get made faster because trust is higher. Teams stop waiting to be told what to do; they know their roles and responsibilities and have the autonomy to own them.

One of the most consistent patterns I see is the jump in communication effectiveness. Leaders who complete our work together report that difficult conversations become less charged. They know how to read a room, manage their own emotional response in real time, and speak in a way their audience can actually receive.

Beyond the individual, I have seen teams go from fractured and underperforming to cohesive and results-oriented, not because the strategy changed but because the person at the top changed how they showed up. Leadership is the variable that moves everything else.

3. Can you share a challenge you faced and the results that came from overcoming it?

The hardest challenge I have faced in this work is convincing high performers that their emotional intelligence requires the same investment as their technical expertise. The executives who need this work most are often the ones most resistant to it, because everything in their career to that point has told them that results are enough.

Early in my practice, I sat across from a founder who had built an impressive company and was losing his best people one by one. He was brilliant, relentless, and completely unwilling to see that his leadership style was the common thread. Getting him to look at that honestly, without it becoming an attack on his identity, required an outside voice, patience, and a different kind of conversation than he had ever been in.

What came out the other side was a leader who not only retained his core team but rebuilt trust that had been quietly fracturing for years. That experience shaped how I approach the work now. The resistance is often the signal that the work is needed most, and I learned to walk toward it rather than around it.

4. What does innovation look like in your day-to-day work?

In my work, innovation means engaging with people as they are, rather than according to standard coaching frameworks. Every leader brings a different set of experiences, blind spots, history, and pressures. A methodology that worked for a corporate VP may not translate to a first-generation founder navigating both personal and professional stakes at once.

On a practical level, it means I am constantly customizing the frameworks I use. The Strategic Transition System I developed came directly from watching patterns in how executives stall when moving from a corporate identity into entrepreneurship. The Emotional Intelligence baseline work in my book grew out of years of noticing where leaders consistently get stuck. I am always reading the gap between what exists and what my clients actually need.

Day-to-day, innovation shows up as adaptability. I do not walk into an engagement with a fixed script. I walk in with curiosity, a set of tools, and the experience of knowing which one the moment calls for.

5. How do you approach innovation in spaces that resist change?

Resistance to change is not a problem to eliminate. It is information. When people push back, they are telling you something true about what is at stake for them, and a good leader or advisor listens before pushing forward.

The biggest mistake organizations make is rolling out changes without accounting for the costs to the people implementing it, at least in the short term. Learning a new process takes time. Shifting a workflow creates friction before it creates efficiency. When leadership skips over that reality, resistance hardens into something that not only costs effectiveness but also an opportunity to build trust with those affected. That missed opportunity often ends up costing more than the original change.

My approach is to bring the people most affected by a change into the conversation as early as possible. Not to let them veto the decision, but to let their voice of experience be heard and give them a real role in shaping how change happens. That is a form of respect, and respect is what makes adoption possible.

Implementing change by making decisions behind closed doors without input tells people that their experience does not matter. There are many better ways to manage change in an organization, and the leader who gets this right has an easier time implementing meaningful changes.

6. How do you develop a strategy in times of uncertainty or rapid change?

We are no longer in a moment where uncertainty is the exception. It is the operating condition. With the pandemic, global instability, economic volatility, and technological disruption, there will always be another thing around the corner that nobody saw coming. The leaders who build a strategy around solving the current problem are always a step behind.

An effective strategy focuses on understanding what people need to navigate disruption successfully. Each crisis impacts individuals in unique ways. Leaders who know their teams deeply can shift quickly because they are not guessing what their people need.

I often say that leaders who chase problem solutions first end up looking like six-year-olds on a soccer field, running toward the ball wherever it is. Effective leaders focus on creating stability so problem-solvers can think clearly. That is a completely different orientation, and it is the one that produces durable results.

7. What role does data, technology, or AI play in your strategic decisions?

AI is a tool that changes the speed and scale of what is possible, and like any tool, what matters is the judgment of the person using it. When leaders use AI to research faster, synthesize information, or stress-test an idea, that is additive. When they use it as a substitute for reading the room, they lose trust, and that is something they cannot afford to lose.

The leaders I work with are often under pressure to do more with less, and AI can look like the answer to that problem. But shortcutting the human dimension of leadership with automation is a false sense of security. If your team cannot hear your voice in a communication, you have already paid a price that no efficiency gain covers.

I use technology in my own practice, including AI for research and development. The information I receive is always processed through my own insight, and delivery to a client is calibrated to what I understand that specific person needs through our work together. Every leader needs to make the distinction between the right role for AI in their circumstances, and, once done, be vigilant in its use.

8. How has your leadership or thinking evolved over time?

Early in my career, I knew EQ in leadership was important. I might not have named it EQ exactly, but I was seeing patterns everywhere I looked. What I found obvious was completely unnoticed by my colleagues.

I recognized the differences each leader made when they followed through on, or didn't, by changing how they showed up. I watched some leaders gain amazing respect and productivity from their teams, and, unfortunately, I saw others implode because they couldn't accept that they were the root of their problems.

Every moment of every day, leaders make decisions, communicate with various audiences, and shape how their company runs day to day. Each one of these tasks is better served by an emotionally intelligent leader.

My work has changed in how I present EQ. Whenever possible, I would incorporate it during business advising sessions. Now I understand this is the basis of success as a leader in today's environment. This is no longer a *"nice to have"* skill; it is an essential business skill to build and scale.

9. What message or insight do you want IRIS readers to take away from your story?

That emotional intelligence is not a luxury for the people who have already made it. It is the infrastructure that gets you there and keeps you there.

The technical skills that open the door to leadership are rarely what determine how far you go or what you build once you are inside.

What determines that is your ability to know yourself, lead, and create the conditions in which the people around you can do their best work. That is EQ in practice.

And for the founders and executives reading this who are already performing at a high level, my message is specific: the ceiling you are hitting is likely not a strategy problem or a market problem. It is a leadership problem.

That's good news, since leadership can be developed.

The investment in your emotional intelligence is not about becoming someone different. It is about becoming a more complete version of who you already are. That version is more effective, more influential, and frankly, more you.

DR. JOSIAS JEAN-PIERRE

Dr. Josias Jean-Pierre is an acclaimed author, speaker, and leader with national and international influence.

He is among the global thought leaders shaping conversations about purpose, leadership, and personal development through his inspiring voice and message.

As an award-winning author and speaker, Dr. Jean-Pierre has been featured in Sheen Magazine, appearing on the New York Times Square Billboard, and in Los Angeles Wire, sharing the spotlight with global entertainers such as Ariana Grande and Lil Nas X.

His achievements include features in Yahoo Finance alongside Ice Cube and recognition in prestigious business publications: Business Elite 40 Under 40, Puget Sound Business Journal 40 Under 40, Forbes 40 Under 40, and Forbes One. Voice of Change Magazine named him one of 40 speakers shaping the future.

Dr. Jean-Pierre's international honors include the Nelson Mandela Award (UK) for distinguished community impact, the ChristLit Book Award (UK) for literary excellence, recognition as Author of the Year in Durban, South Africa, and selection among the *"100 Prolific Personalities"* in Johannesburg, South Africa, celebrating his inspirational leadership.

Author of eight books on faith, purpose, leadership, and personal growth, Dr. Jean-Pierre is dedicated to helping people turn pain into purpose and challenges into opportunities.

Motivated by a sense of calling and service, Dr. Jean-Pierre's mission is to empower people to realize their potential, overcome limitations, and live purposefully.

1. The impact I am most proud of is witnessing individuals achieve positive transformations that empower them and future generations. Creating lasting change starts with the right tools and mindset, enabling people to shape a better tomorrow.

2. ***Who or what has been most affected by your work?***

My work has had the greatest effect on individuals who have felt overlooked or ready to give up. Because of my efforts, these individuals gain the confidence to recognize their own worth and potential.

3. ***Meaningful impact*** in my industry means reaching one, teaching one. Though the impact has touched many, my focus is on everyone. If I reach one, they reach another. Each one reaches one, each one teach one.

4. ***What results best reflect your progress or success?***

The result that best reflects my success is international recognition: being named an icon in Africa and receiving awards from the U.K. These honors show that my message has a positive impact across diverse regions worldwide.

5. ***I overcame doubt from others and chose self-belief;*** this led to three presidential lifetime achievement awards, being named the USA's most influential motivational speaker for 2026 by Conglomerate Magazine, and inclusion among 40 speakers shaping the future with Les Brown in Voice of Change magazine.

6. ***I stay focused by returning to my why, which fuels my work and keeps my mission clear.***

7. ***Innovation*** is the practical transformation of ideas into genuine, measurable value not just new, but useful and impactful.

8. ***What does a strong strategy look like when it's executed well?***

When executed well, strategy becomes an ongoing organizational discipline that seamlessly bridges vision and daily actions.

STACEY D.M. SHAW

Stacey D.M. Shaw is a dynamic force in the business world; a multifaceted, self-motivated individual with over 25 years of C-suite administrative support within several industries.

Her diverse background encompasses facilities management, business ownership, real estate, and construction support. Her experience shaped her path to becoming a Virtual Chief Administrative Officer and Operations Strategist, helping her clients' operations run more efficiently and productively.

An advocate for women in transition and for women who experience abuse, Stacey's mission is to help educate and empower women to live a life of freedom after abuse.

She currently serves as the Founder, Virtual Chief Administrative Officer, and Operations.

Strategist for Your Brilliant Transitions, an administrative consulting company; AdvisoryBoard Member for ReDevelop Her - a 501c3 organization that assists divorced women; and advisory board member for Senator Ivy's Domestic Violence Committee. Founder of the women's empowerment group - Her Chosen Legacy; The Her Chosen Legacy Summit and Magazine.

Stacey's journey is filled with accomplishments a Christian Biblical Life coaching certification in 2010, a Real Estate License in 2019, and a Soul-Care Coaching Certification in 2020. She is a contributing author in three published anthologies, "Surviving Her: Heal Your Soul... Heal Your Business and "The Beautiful Face of The Invisible Illness; Phenomenal Woman Award Recipient in 2023 and 2024; Speak Life Award Recipient 2023 and 2024; 2024 Passion Purpose Peace Award and cover features of My Sister Keeper April 2023 Magazine.

Recently, she wrote her own book, Administrative Foundations for Success:40 Administrative.

Tips Every Small Business Owner Needs

Her years of experience, training, and decision to live life with total intention have helped her walk in her purpose.

However, her greatest joy and pleasure are serving the Lord, being a mother to her two wonderful children, exploring nature, enjoying a full day at the beach, and listening to and dancing to music.

Stacey is more than a name; she is a symbol of empowerment, a testament to the boundless possibilities of living with intention. Get to know her and the silly, fun-loving individual she can be. As her tagline says, she is: ***`Jesus' Baby Girl'.***

Stacey D.M. Shaw

Chief Administrative Strategist- The Administrative Architect
Your Brilliant Transitions, LLC

URL: yourbrillianttransitions.com

What impact are you most proud of creating through your work, and why does it matter?

The impact I am most proud of is helping business owners bring structure and order to their internal operations. Many businesses grow by focusing on the vision, services, and opportunities ahead, yet the operational side of the business often hasn't been designed to support that growth.

My work focuses on building the administrative infrastructure that allows a business to function with clarity, accountability, and operational support. This includes systems, documented processes, administrative structure, and the intelligent use of technology, including AI, to strengthen how the business operates.

This matters because when the internal side of a business is structured correctly, business owners regain time, teams gain direction, and the business can grow without constant operational disruption.

Who or what has been most affected by your work?

The people most affected by my work are business owners and entrepreneurs who have reached a stage where growth has exposed gaps in their internal operations.

Many of them are successful and expanding, yet they are shouldering operational responsibilities that should be handled by systems and support structures.

When the internal structure of the business is strengthened, everything begins to shift. Business owners can focus on strategic direction, teams operate with greater coordination, and the business begins to function as a cohesive system rather than a collection of disconnected tasks.

How do you define meaningful impact in your industry?

The effectiveness of administrative operations becomes evident when a business continues to run smoothly even without the owner directly overseeing every aspect.

When processes are documented, responsibilities are clear, and operational systems are in place, the business becomes stable and capable of sustained growth. Administrative infrastructure enables the business to operate consistently rather than relying on constant oversight.

Impact is reflected in operational stability, stronger execution, and the business's ability to grow with structure and support.

What results best reflect your progress or success?

The strongest results appear when business owners begin operating with greater clarity and confidence. They delegate tasks efficiently, team roles are clear, and operations run smoothly.

Another powerful indicator of progress is when a business no longer depends on one person to manage every moving part. When administrative systems, processes, and operational frameworks are in place, the business gains a structure that supports both stability and expansion. Business owners can then focus on income-producing activities, strategic direction, and client growth.

Can you share a specific obstacle and the results that followed?

One of the most common obstacles I encounter is rapid business growth without the operational systems or administrative foundation to support it.

Although business owners and entrepreneurs are putting in significant effort, they still face challenges such as miscommunication, ambiguous roles, a lack of documented processes, and minimal administrative systems to manage workflow.

Through structured administrative assessments and operational restructuring, we begin identifying the gaps. From there, we implement systems, documented processes, clearer workflows and structures, and in some cases, AI-supported tools that strengthen how the business operates.

Once those elements are in place, coordination improves, team members work with greater clarity, and business owners can redirect their attention toward strategy, revenue generation, and long-term growth.

How do you stay focused on results without losing sight of your mission?

My mission is to help business owners create solid operational structures for stable, clear growth.

Results are simply evidence that the structure is working. When systems are in place, processes are documented, and administrative foundations are established, the business begins operating more effectively.

By consistently focusing on operational clarity, accountability, and strong support systems, my work helps businesses move forward sustainably.

What does innovation mean to you beyond buzzwords?

Innovation is often associated with adopting new technology, yet real innovation is about improving how a business operates.

In administrative operations, innovation manifests in stronger systems, improved processes, more robust communication frameworks, and the thoughtful integration of tools such as artificial intelligence. AI is becoming an important support tool, helping businesses organize information, automate repetitive tasks, and strengthen operational decision-making.

True innovation refines how workflows are managed through a business, enabling business owners and their teams to operate with greater efficiency, structure, and clarity.

How do you encourage innovation within your organization or work?

Innovation begins with structure. When processes, responsibilities, and systems are clear, people have the confidence to identify opportunities for improvement and contribute ideas that strengthen the business.

By establishing strong operational frameworks and administrative support systems, I help create environments in which businesses can continuously improve the way work is executed. This includes reviewing workflows, strengthening systems, building and executing a solid administrative foundation, and exploring how emerging tools, including AI, can support the business in practical ways.

How are emerging technologies, including AI, influencing your approach?

Artificial intelligence is beginning to reshape how administrative work is performed and how businesses analyze operational information.

AI helps businesses automate routine tasks, organize data efficiently, and gain operational insights.

In my work, AI is not used to replace business owners. It is used to strengthen the administrative infrastructure. When implemented thoughtfully, it supports business owners and their teams, enabling them to focus more on strategy, client experience, and business growth.

What does a strong strategy look like when it is executed well?

A strong strategy becomes visible through execution. Everyone involved in the business understands the direction, responsibilities are clearly defined, and systems are in place to support progress.

When strategy is executed well, business owners spend less time managing operational problems and more time guiding the business's future. Clear systems, aligned teams, and disciplined execution allow strategy to move from vision to measurable results.

Made in United States
North Haven, CT
04 November 2026

21734480R00027